The Science of Bakery Products

The Science of Bakery Products

Edited by

W. P. Edwards
Bardfield Consultants, Braintree, Essex, UK

ISBN: 978-0-85404-486-3

A catalogue record for this book is available from the British Library

Published by The Royal Society of Chemistry,
Thomas Graham House, Science Park, Milton Road,
Cambridge CB4 0WF, UK

Registered Charity Number 207890

For further information see our web site at www.rsc.org

Preface

This book has its origins from when I moved from the confectionery industry to the flour milling business. One of the reasons I wrote this book was because I failed to find a suitable book on the science and technology of baking. It is intended to be used along with my Science of Sugar Confectionery book in the RSC Paperbacks series. The book is aimed to fulfil the needs of students at A level and above who need to know the science of baking. It also intends to cover the needs of newcomers to the baking industry.

Making baked products, particularly bread, is one of the oldest human activities. This book attempts to explain the underlying science behind making baked products. It is aimed at both scientists and science students for understanding the application of science and technology of baked products, as well as bakers and apprentices who want to understand the science.

There are two important decisions that the author of a work like this has to make – the topics to cover and the topics to omit. These decisions have been made on the basis of what information is likely to be useful to the reader.

W.P. Edwards
Bardfield Consultants, Braintree, Essex, UK

Contents

CHAPTER 1

Introduction

1.1 HISTORY

Baking, particularly the baking of bread, is one of the oldest of human activities – indeed one of the oldest surviving papyri appears to be a set of instructions for making bread. Another document is part of a correspondence explaining that pyramid construction is falling behind because the supply of beer and bread to the labourers has been insufficient, thus revealing that the diet of labourers has changed relatively little in thousands of years.

Western civilisation is based on the cultivation of wheat, a practice that seems to have started in Mesopotamia, the area that is currently Iraq. Wheat is a member of the Grammacidae, *i.e.* it is a member of the grass family. The cultivation of wheat spread from the Middle East across Europe. Settlers took wheat seeds with them to the Americas and started to cultivate wheat there. Those settlers from Great Britain took wheat that had evolved to grow in British conditions. These wheat varieties would grow on the eastern seaboard but were not successful in the American Midwest. Subsequently, however, wheat from Eastern and Central Europe was found to thrive in the Midwest. The cultivation of wheat also spread to Canada and Australia.

In Great Britain, the industrial revolution in the nineteenth century was initially good for the farming community – as people moved from subsistence agriculture to the factories it created markets for agricultural products. This situation continued up to the 1880s when quantities of imported grain started to become available. This imported grain was much harder than English wheat and created a problem since the wind and watermills could not grind it. A solution appeared in the form of the roller mill, a Hungarian invention, which could cope with hard North American wheat. These roller mills could easily produce much whiter flour than the old stone mills. The large milling companies set up mills on dockside sites as the most economic way of handling imported grain. The large wind and water mills that had supplied the cities started to close as they could not compete with these new dockside roller mills. Small rural mills, though, continued to trade locally.

The imported grain triggered a farming recession that ran from the 1880s to 1939. British governments became aware of the strategic problems caused by relying on imported food. Research on wheat breeding led to varieties of wheat with good bread making potential that would grow in the British climate. Other research led to the Chorleywood Bread Process that was intended, among other things, to reduce the dependence on imported wheat.

The next stage was Britain's accession to the European Economic Community [EEC, commonly known then as the Common Market, and now known as the European Union (EU)], which meant that the Common Agricultural Policy (CAP) applied. The policy in the form then current sought to penalise the use of food crops from outside the EEC when the crop could be produced inside the EEC.

Originally, the policy had sought to support small farmers by guaranteeing a stable high price for their products. When the supply of a commodity exceeded the demand the surplus was bought and placed in store. This process was called intervention. Keeping stocks of intervention wheat was easier than some other commodities as neither refrigeration nor freezing was needed, unlike the position for butter and beef. There was also the distinct possibility that a bad harvest would allow the grain to be brought out of intervention. The other way of disposing of intervention wheat was subsidised sales on the world market. This was the feature that the traditional wheat-exporting nations objected to most strongly.

If the EEC price was higher than the world price, which it usually was, then imports from outside the EEC had to pay a levy. This provided strong financial motive to try and move from using Canadian wheat. The British wheat that was mainly used instead was not, and is not, as fundamentally well suited to making bread by a long process. Thus, although there were other issues in the move to shorter processes for bread making, the CAP supplied a push because it provided financial advantages for using EEC wheat.

At the time of writing, the World Trade Organisation is pushing for the abolition of agricultural subsidies. If this happens, wheat imported into the EU will no longer be at a financial disadvantage. However, the baking industry is most unlikely to shift back to longer bread making processes. The one area where the use of long processes for bread making has increased is in domestic bread machines, which have increased domestic bread production markedly. Because these machines use a fairly long process the demand for very strong bread flour sold retail has also increased markedly.

The baking industry is not just concerned with the production of bread, there is an important difference between bread and other baked products. Bread is regarded as a staple food and as such attracts regulation of its composition and sometimes price. Biscuits, cakes, pastries and pies are regarded as discretionary purchases and avoid regulation. Bread production is an extremely competitive business while the production of other baked goods is not quite so competitive.

Some supermarkets use cheap bread to attract customers. The supermarket management, knowing that bread is a basic necessity, reckon that if the customer can be lured into the supermarket with the offer of cheap bread then their trade can be captured. Producing the cheapest possible bread does not have a positive effect on quality.

Some small bakers use a variant of the same trick by arranging the shop so that customers need to queue for bread in front of a display of cakes and pastries – which is intended to produce impulse sales.

Another modern trend is the increased sale of filled rolls and pre-packed sandwiches. The sale of filled rolls provides many small bakers with a very satisfactory source of profit. The manufacture of pre-packed sandwiches is now a large industry, consuming large quantities of bread. Such sales growth is obviously caused by a population that is short of time rather than money.

The two hardest decisions in writing this book are what to put in and what to leave out. While the length is decided by the publisher there is no room to produce an encyclopaedia. An attempt has been made to cover examples of the commonest types of product. Inevitably there has to be a great deal about bread in this work but it is not solely about bread, other baked goods have their place in the book just as they do in the bakery industry.

In deciding what to put in and what to leave out, preference has been given to items that are thought likely to be useful to the reader or give an understanding of the current situation. This leads to information on nutrition being included while the genetics of yeast have been left out.

It is a sign of the times that information on nutrition has been included; if this work had been written some years ago it is doubtful if information on nutrition would have been included. At the time of writing there is considerable pressure on the food industry over the unhealthy diet of the general population. The major dietary problem of the western world at present is a diet with too much energy and, particularly, too much fat and salt. The excess energy might be explained by changes in working life as employment becomes less physical and more sedentary. A diet that would be entirely satisfactory for a manual labourer contains far too much energy for an office worker.

1.2 LANGUAGE AND UNITS

The language used will be the Queen's English or that subset of it as approved by the Royal Society of Chemistry (RSC). Where chemical names are concerned there are some lost causes, such as caustic soda, where little would be gained if those who clean factories called this substance sodium hydroxide. Arguably, the name caustic soda conveys more useful information. Similar lost causes are spirits of wine (ethyl alcohol or ethanol) and spirits of salts (hydrochloric acid). While lipid chemists may insist on referring to triacylglycerols many people in industry continue to refer to triglycerides. Similarly trivial names for fatty acids such as lauric will continue to be used. The principle in all of this is to use the proper name but to mention other names that are in common use.

There are a few areas in the subjects covered by this book where unfortunately the same words or are used to describe different things. They are gluten and flour improver. In food law a flour improver would cover a substance added to flour to improve its performance, usually in bread. Such a substance is ascorbic acid. In a bakery, the expression flour improver covers a mixture that is added to the dough. In this context a flour improver will contain not only substances like ascorbic acid but also, for example, enzyme active soya flour, emulsifiers and possibly fat.

Gluten is used to refer to dried wheat gluten, all the proteins in a dough, and by chemists as a class of proteins.

The names given to products can also cause problems. The worst possible case is the English and the American muffin, where the same word is used to describe two completely different products! Product names are a problem not only between the farthest flung parts of the English speaking world but also within the UK. Products that could be classified as a bread roll are known as a breadcake in Yorkshire, a cob in the Midlands, and a huffer in Essex.

These differences occur because there is not a national market in baked products. Short life products of low value have to be made near to where they are consumed, otherwise the cost of transport will exceed the value of the product. In some cases products are not known away from their local market. Such a product is the Scots morning roll. Other products such as Eccles cakes and Chorley cakes are still associated with their area of origin.

While there is a world market for grain and flour there is not a world market for low value short life baked products. The only sort of baked products that can be traded internationally are long life products such as biscuits and rich fruit cakes.

When it comes to units of measurement that subset of the metric system known as SI should be used. The three base units that are relevant here are the metre, kilo and second (replacing the earlier system based on the centimetre, gram and second, *i.e.* cgs units). In practice, some cgs units continue in use. While the UK has officially gone metric the USA continues to use "English Units". Confusingly, some of these units, although they have the same name, are not the same size as the units of the Imperial system used in Britain. The units that cause trouble tend to be the pint, the gallon, and the ton. The Imperial pint is 20 fluid ounces while the US pint is 16 fluid ounces, which leads to the discrepancy between the US ton and the Imperial ton. Curiously, the Imperial ton at 2240 lbs is nearly the same weight as the metric tonne at 1000 kg (= 2120 lbs). In this work, tonne means the metric tonne; the other two tons will not be used.

There is one other set of units: the bag and the sack. Recipes were often classified by so many bags or sacks, meaning bags or sacks of flour. The two should not be confused as a sack of flour held 20 stone of flour, *i.e.* 280 lbs, while a bag held half a hundredweight, *i.e.* 56 lbs. Modern British bags of flour normally hold either 25 or 32 kg. Flour for domestic use normally comes in 1.5 kg bags.

1.3 FOOD LAW

Legislation has its effects on all parts of the food industry and anyone working in the food industry needs to be aware of this. Although both science and the law try to be exact there are differences between the scientific and legal approaches. In particular, the use of language is different. In the "Bread and Flour Regulations 1998", for example, there is not unreasonably a definition of bread which is that "bread" means

> a food of any size, shape or form which – (a) is usually known as bread, and (b) consists of a dough made from flour and water, with or without other ingredients, which has been fermented or otherwise leavened and subsequently baked or partly baked, but does not include buns, bunloaves, chapatis, chollas, pitta bread, potato bread or bread specially prepared for coeliac sufferers.

It is obviously entirely reasonable that bread specially prepared for coeliac sufferers, *i.e.* those allergic to gluten, should be outside the regulations. Chapatis are clearly outside the regulations anyway as they are unleavened but chollas and pitta bread are clearly bread in common parlance. The case of pitta bread is interesting as the regulations specifically exclude it from the category of bread yet it is called bread. This is a product that in common parlance is bread, but is not bread within the scope of the Bread and Flour Regulations 1998.

In Great Britain, modern food law developed from the Food and Drugs Acts. This law came about after an outbreak of arsenic poisoning among beer drinkers. The cause of the problem turned out to be glucose that had been used in making the beer. The glucose had been made by hydrolysing starch with sulfuric acid. In turn, the acid had been made by the lead chamber process from iron pyrites, which contained arsenic as an impurity. The approach adopted was that all foods should be "of the substance and quality demanded". This was obviously intended to cover any future problems with some other contamination, not necessarily with arsenic.

The arsenic in beer incident has led to the rule that all food ingredients must be food grade, as must any food contact materials or materials used in the process. This rule can have the consequence that a material such as coke must be food grade. If a food is to be roasted over coke then that coke must be food grade. A non-food grade coke might, possibly, contain a substance such as arsenic.

The only food law prior to this time was the statutes of bread and ale. These set out to regulate the brewing of ale and the quality of bread. As no modern analytical methods were available the beer was assessed by pouring some on the end of a barrel and an official, called the ale conner, sat on the beer in leather breeches. If the breaches stuck to the beer the ale failed. Presumably, the test detected residual sugar. Bread was tested by examining loaves, probably to detect contamination with the fungus known as rope. A baker whose loaf failed had the offending loaf strung round their neck followed by being whipped around the town or stood in the pillory. Other countries, particularly those whose legal systems follow Roman rather than Anglo-Saxon law, have tended to more prescriptive laws.

The British approach is to allow any ingredient that is not poisonous unless the ingredient is banned. Additives are regulated by a positive list approach. Unless the substance is on the permitted list it can not be used. There are anomalies where a substance can be legal in foods but not be permitted to be described in a particular way. An example is the substance glycherrzin. This is naturally present in liquorice and has a sweet flavour. It would be illegal to describe it as a sweetener as it not permitted as a sweetener. The substance is permitted as a flavouring, however, and can be added to a product. This will make the product taste sweeter than it would without the addition. Conversely, the protein thaumatin is permitted as an intense sweetener. In practice, thaumatin has more potential as a flavouring agent. It would have been much easier and cheaper to obtain approval for thaumatin as a flavouring than as a sweetener.

The British system does not automatically give approval to ingredients merely because the ingredient is natural. This is in contrast to the position in some other countries. There will always be grey areas. One example is the position of the oligo-fructose polymers that are naturally present in chicory. Chicory is undoubtedly a traditional food ingredient; however, oligo-fructoses extracted from it are not necessarily a traditional food ingredient. If the fructose polymers are hydrolysed to fructose then that is a permitted food ingredient. However, if they are partially hydrolysed what is the status of the resulting product? The issue of fructose polymers is further complicated because one of their interesting properties is that they might not be completely metabolised. If that is the case then they would be considered as additives rather than ingredients. Additives need specific approval while ingredients do not.

1.3.1 Bread and Food Law

There are exceptions to the general British approach that what is not illegal is legal and the most important of these is bread. Bread as a staple food has attracted regulation about its composition. Governments have taken the view that while the nutritional composition of minor components of the diet do not matter, deficiencies in the composition of bread will have a significant effect on the national diet. Various countries take different views on the level of fortification necessary in bread. The USA has bread regulations that call for more fortification than the British regulations do, while the French regulations call for less fortification of bread than either. It has been suggested that there should be a minimum level of selenium added to British bread and flour because most of the nation are now deficient in selenium. This deficiency is caused by the switch from bread made from North American wheat to that made from EU wheat. North American soils contain more selenium because they are geologically younger than European soils which have had sufficient time for the selenium to wash away. Selenium fortification could be achieved by adding selenium to fertiliser, as is done in Finland, rather than to flour. The dosing regime for selenium would need to be very carefully controlled as an excess would be poisonous. Fortification of chapatti flour with vitamin A has been proposed to prevent rickets in Asian girls but fortification with fat-soluble vitamins is inherently risky as an overdose could be toxic since the body is not well equipped to dispose of an excess.

Bread and flour are specifically regulated in the UK, at the time of writing, by the Bread and Flour Regulations 1998, The Food Labelling (Amendment) regulations 1998, and the Miscellaneous Food Additives (Amendment) Regulations 1999.

Current British fortification of bread and flour is restricted to fortifying white and brown flour and bread with those materials that would be present in wholemeal bread or flour.

White and brown flour in the UK have to be fortified with calcium, iron, thiamine (vitamin B1), and niacin (vitamin B3; also known as nicotinic acid) as follows: flour should contain not less than 235 mg per 100 g and not more than 390 mg per 100 g calcium carbonate, iron not less than 1.65 mg, thiamine (vitamin B1) not less than 0.24 mg, and nicotinic acid not less than 1.6 mg or nicotinamide (which is nutritionally equivalent to nicotinic acid) not less than 1.60 mg.

The regulations specify which flour treatments can be used in bread and flour and in which circumstances they can be used (Table 1). One obvious anomaly is that ascorbic acid is permitted in wholemeal bread but not in wholemeal flour.

Miscellaneous Food Additives (Amendment) Regulations 1999 lays down the levels of propionates that can be added to bread. It also notes that propionates can form naturally in the fermentation of bread.

The EU has sought over the years to harmonise the regulations governing food products as an aid to free trade across the EU. In the case of chocolate this has taken 30 years! Currently, the approval of novel ingredients has moved to the EU, with national governments passing the necessary legislation afterwards. This avoids the previous situation where new ingredients were assessed by the member states individually.

One of the stranger EU compromises is the "Legal in one country" rule. This solves the problem of trade across the EU by stating that if a product is legal in one EU state it can be sold throughout the EU, even in states where the product is illegal. This obviously ends up with the position that a product can be sold in a country where it can not be made legally.

There has been relatively little problem with baked goods as the shelf life inhibits cross border trade. The one area where the regulations were

Table 1 *Permitted ingredients in flour and bread*[a]

Ingredient	*E number*	*Types of flour and bread where it may be used*	*Maximum quantity permitted in mg kg^{-1} of flour*
Sulfur dioxide Sodium metabisulfite	E220 E223	All flour for use in the manufacture of biscuits or pastry except wholemeal	The total quantity of these additives used must not exceed 200 calculated as sulfur dioxide
L-Ascorbic acid	E300	All flour except wholemeal. All bread	200
L-Cysteine hydrochloride	E920	All flour used in the manufacture of biscuits, except wholemeal or flour to which E220 sulfur dioxide or E223 sodium metabisulfite has been added	300
L-Cysteine hydrochloride	E920	Other flour except wholemeal. All bread, except wholemeal	75
Chlorine[b]	E925	All flour intended for use in the manufacture of cakes, except wholemeal	
Chlorine dioxide	E926	All flour except wholemeal. All bread, except wholemeal	30

[a] From The Bread and Flour Regulations 1998 Schedule 3 Regulation 5 "Ingredients Permitted in Flour and Bread", which is Crown copyright.
[b] The use of chlorine has been phased out since November 2000.

different concerns the use of potassium bromate as a flour treatment. Only the UK and Ireland permitted potassium bromate, but its use has since been prohibited.

Differences in the regulations between countries are not always easy to explain. If a substance is harmful it might be thought that it would be harmful everywhere. In some cases substances are not permitted because no one wanted to use them, and hence there was no application to permit them. Other cases, such as that of potassium bromate, are much harder to explain. Essentially, the British authorities were convinced that potassium bromate was both carcinogenic and that traces were present in the finished product. At the time of writing, the US authorities have not been convinced and continue to permit potassium bromate.

1.3.2 Health and Safety

Bakery products are not inherently dangerous but the following points should be made: some bakery products are cooked at high temperatures, *e.g.* 230°C, which is hotter than most forms of cookery even if it not a high temperature by chemical standards.

Precautions must be taken to prevent contact between people and hot equipment or products. There is a particular risk with concentrated sugar-containing syrups or jam as they not only have a high boiling point but they are by nature sticky and a splash will tend to adhere.

Clearly, precautions should be taken to prevent burns and to deal with any that occur. In the event of a burn either plunging the afflicted area into water or holding it under running water is the best first aid. A sensible precaution is to make sure that either running water or a suitable container of water is always available.

Most bakery ingredients are not at high risk of bacterial contamination. However, some ingredients are prone to bacterial problems, *e.g.* flour, egg albumen and some of the gums and gelling agents. In handling these materials precautions need to be taken so that they do not contaminate any finished product or other ingredients. Bakery ingredients should be food grade. Any product being made to be eaten should be produced using food grade equipment and not in chemical laboratories. Precautions do need to be taken so that dusts from handling ingredients do not cause eye or lung irritations. Some bakery ingredients, although perfectly food grade and edible, can cause irritations if inhaled. The commonest example is flour. Another potential hazard with flour in that when handled in bulk it can create the conditions for a dust explosion and fire.

CHAPTER 2

Science

2.1 BASIC SCIENCE

Certain items of science that are fundamental to bakery products are discussed here.

2.1.1 Stability

Some bakery products have a long shelf life while others have a relatively short life. In some products the shelf life is limited not by the bakery component but by the filling, *e.g.* meat pies and cream cakes.

In general, the longevity of long life products is due to their low moisture content, which is too low for spoilage organisms to grow.

2.1.2 The Water Activity

The relevant parameter is not only the water content but the water activity. Water activity is a thermodynamic concept invented to explain why materials with different levels of water content do not behave in the same way chemically or biologically. The water activity reflects the ability of the water to be used in chemical or biological reactions. It is the concentration corrected for the differences in the ability of the water to undertake chemical reactions. If a non-volatile solute were dissolved in water then the vapour pressure should decrease in a certain way for a perfect mixture. A thermodynamically ideal substance would always have a water activity of one. Originally, water activity could only be measured indirectly. One such method that was used in the food industry was to measure the weight loss of the product when held at a range of controlled relative humidities. This had the effect of holding the product over a range of water activities. If the product was held at its own water activity it would neither gain nor loose weight. This water activity was said to be its equilibrium water activity.

2.1.3 The Equilibrium Relative Humidity (ERH)

This term is normally abbreviated to ERH. The ERH is deduced by extrapolating the weight loss values over a range of water activities greater and lesser than that of the product. Where the two lines intersect lies the water activity of the product. This extremely tedious and time consuming method has largely been superseded by instruments that measure the water activity directly. The ERH still has practical importance since it indicates the conditions under which the product can be stored without deterioration.

2.1.4 The Dew Point

A related property is the dew point. This is the point at which condensation occurs on cooling. When products are cooled the temperature must not fall to the dew point or condensation will occur. If condensation occurs on the product then product spoilage is likely.

2.2 COLLIGATIVE PROPERTIES

2.2.1 Boiling Points

Colligative properties are defined as depending on the number of particles present rather than the nature of the particles. In sugar confectionery the most important of these is the elevation of boiling point. Because sugars are very soluble, very large boiling point elevations are produced, *e.g.* as large as 50°C. As the elevation of the boiling point is proportional to the concentration of the solute, the boiling point is, unsurprisingly, used as a measure of the concentration and hence as a process control.

The boiling point of a liquid is the temperature at which its vapour pressure is equal to the atmospheric pressure. If the pressure is increased the boiling point will increase, while reducing the pressure will reduce the boiling point. Sugar syrups are often made by boiling up a mixture of sugars to concentrate them. The use of a vacuum has several advantages. Energy consumption is reduced, browning is reduced and the process is speeded up. A common practice is to boil a mixture of sugars under atmospheric pressure to a given boiling point. A vacuum is then applied. This causes the mixture to boil under reduced pressure. This not only concentrates the mixture, but the latent heat of evaporation cools the mixture rapidly, speeding up the production process since the product will ultimately have to be cooled to ambient temperature.

Another area where boiling points are important is regarding steam. A large proportion of food industry plant is heated by saturated steam, *i.e.* steam at its boiling point. The temperature of steam can be regulated by controlling the pressure. One side-effect of using vacuum boiling rather than boiling at atmospheric pressure is the that lower steam pressures are needed because the boiling point has been reduced. Lower steam pressures produce a considerable saving in terms of the capital cost of steam boilers and pipe work since they do not have to be built to stand the higher pressures.

2.2.2 Measuring Vacuum

In controlling a process the level of vacuum obtained controls the amount of water in the product, which is obviously important. The level of vacuum applied can be measured in several ways. Mercury manometers are not now used although they may once have been. The commonest measuring instrument is probably the Bourdon gauge, although various designs of pressure sensor are also used. The gauge could be calibrated in several different units. Calibrations in units of length, *e.g.* inches or millimetres of mercury, are common – a legacy of using a mercury manometer. Alternative units of pressure used are pounds per square inch or newtons per square metre. Another system uses bars or millibars (where one bar equals one atmosphere).

2.3 pH

The pH scale is a convenient way of measuring acidity or alkalinity. The definition is: $\mathrm{pH} = -\log_{10}[\mathrm{H}^+]$.

This has the considerable advantage that it almost always gives a positive number. On the pH scale 14 is strongly alkaline whilst 1 is strongly acid. The pH system implies that the solution is aqueous. When, as not infrequently happens in sugar confectionery, there is a higher concentration of sugar than water it does imply interesting questions regarding the result produced by a pH probe.

In bakery products the pH of the product is important for several reasons. Acid foods are normally relatively safe microbiologically as a low pH inhibits the growth of bacteria. Fruit flavoured products such as fruit pie fillings normally have some acid component added to complement the fruit flavour. Where a hydrocolloid is present the pH of the product can be critical in terms of product stability or gelling. A hydrocolloid held at its isoelectric point, *i.e.* the pH at which there is no net charge, will likely come out of solution.

2.4 POLARIMETRY

When illuminated, any molecule possessing an asymmetric centre will rotate the plane of plane-polarised light. Most sugar confectionery ingredients are optically active. By measuring the optical rotation of solution the concentration of sucrose or other sugars can be measured. When sucrose is broken into fructose and dextrose the rotation of polarised light is reversed. Hence this mixture of sugars is normally referred to as invert sugar. In confectionery factories polarimeters are used to check the concentration of products and components. This is not a particularly accurate practice but it suffices.

2.5 THE MAILLARD REACTION

Maillard reactions are non-enzymic browning reactions. In practice any browning in foods is a Maillard reaction, except where it is enzymic, *e.g.* the browning of a cut apple is enzymic and, hence, it is not a Maillard reaction. The Maillard reaction is not a name reaction whose details can be found in a text book. The term covers a whole range of reactions that occur in systems ranging from food to the life sciences. In sugar confectionery, the problems with Maillard reactions lie in preventing them where they are not wanted, *e.g.* in boiled sweets, and in encouraging them where they are wanted, *e.g.* in toffees. The name of the reaction goes back to Louis Camille Maillard, who heated amino acids in a solution with high levels of glucose.[1] The chemistry of the Maillard reaction is easily described as complex. It is complex not only because the reaction can give complex products but the starting materials are themselves complex. Most model systems involve studies of one reducing sugar being heated with one amino acid. A typical confectionery system like a toffee would involve heating a mixture of proteins, usually from milk with a mixture of reducing sugars and fats. In sugar confectionery the conditions of the reaction are likely to be high temperature but low water activity. In the early stages of the reaction the free amino group of an amino acid, usually in a protein, condenses with the carbonyl group of a reducing sugar. The resulting Schiff bases can rearrange by Amadori or Heyns rearrangement to give an N-substituted glycosylamine that can degrade to fission products by a free radical mechanism. In the advanced stages of the reaction the Amadori or Heyns rearrangement products degrade by one of three possible routes: via deoxysones, fission or Strecker degradation. Figures 1–5 show these reaction schemes. The 1-deoxyglycosones and 3-deoxyglycosones can form reactive α-dicarbonyl compounds such as pyruvaldehyde and

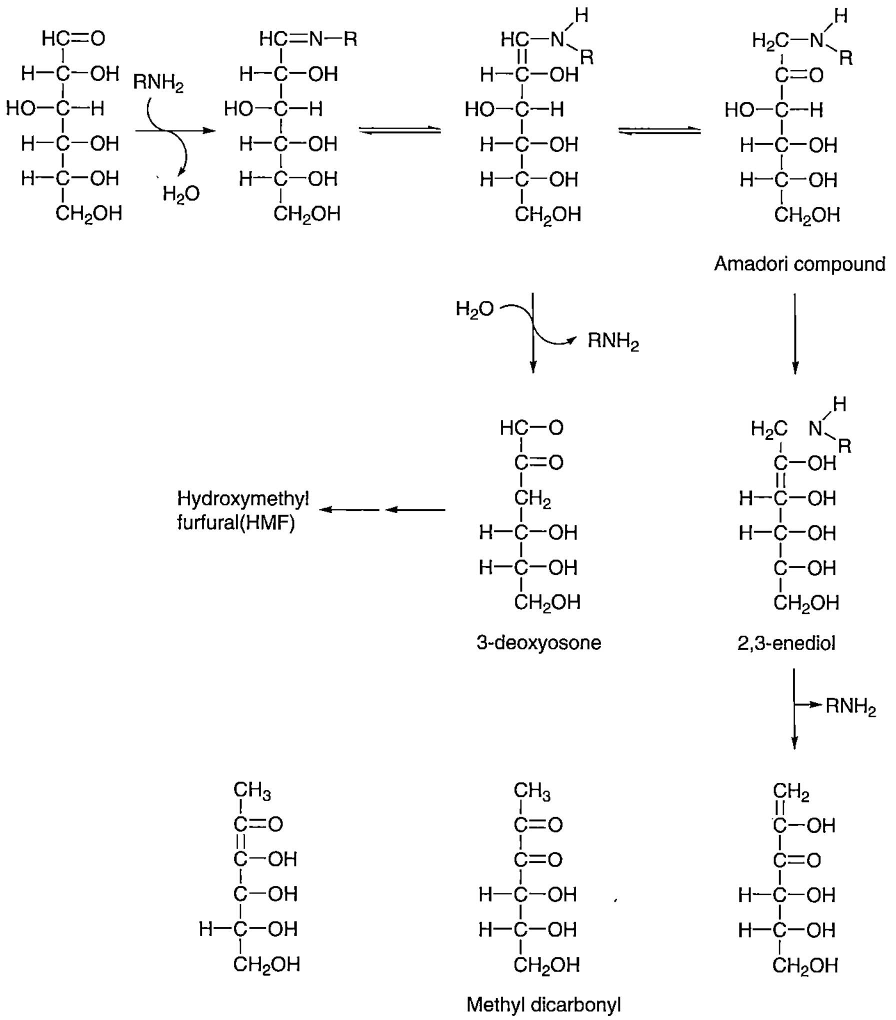
RNH2
H2O
Amadori compound
RNH2
Hydroxymethyl furfural(HMF)
3-deoxyosone
2,3-enediol
RNH2
Methyl dicarbonyl

Figure 1 *Part of the Maillard reaction*

diacetyl by a retro-aldisation reaction. These reactive intermediates are then available to react with ammonia and hydrogen sulfide.

2.5.1 Sulfur-containing Amino Acids

While sulfur-free amino acids are broken down to amines via decarboxylation, the sulfur-containing amino acids such as cysteine can undergo more complex reactions. Fisher and Scott suggest that, because cysteine produces a powerful reducing aminoketone, hydrogen sulfide could be produced by reducing mercaptoacetaldehyde or cysteine.[2]

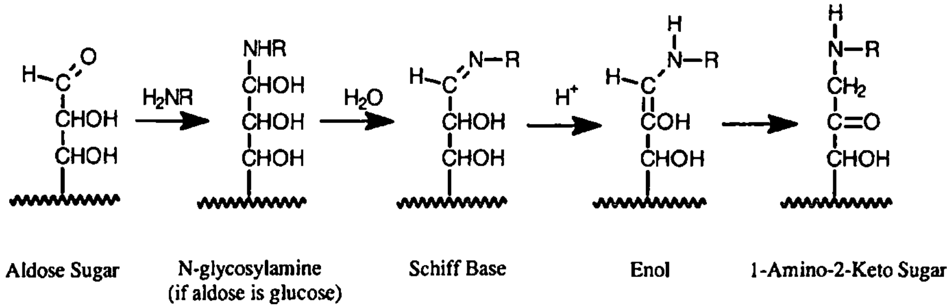

Aldose Sugar | N-glycosylamine (if aldose is glucose) | Schiff Base | Enol | 1-Amino-2-Keto Sugar

Figure 2 *Formation of 1-amino-2-keto sugars from aldose sugar*
(Reprinted with permission from *Food Flavours: Biology and Chemistry*, C. Fisher and T. R. Scott, ©1997 Royal Society of Chemistry)

Ketose Sugar | N-Fructosylamine (if ketose is fructose) | Enol | 2-Amino-1-Keto Sugar

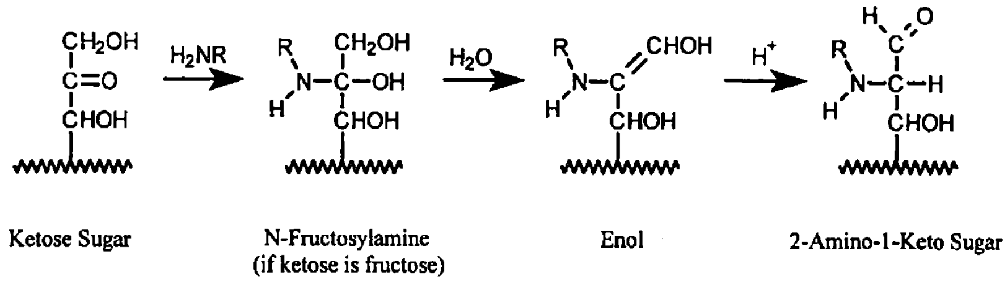

Figure 3 *Formation of 2-amino-1-keto sugars from ketose sugar*
(Reprinted with permission from *Food Flavours: Biology and Chemistry*, C. Fisher and T. R. Scott, ©1997 Royal Society of Chemistry)

Alternatively, hydrogen sulfide could be produced alongside ammonia and acetaldehyde by the breakdown of the mercaptoimino-enol intermediate of the decarboxylation reaction of the cysteine-dicarbonyl condensation product. Fisher also points out that hydrogen sulfide is forms many odiferous an hence intensely flavoured products.[2] Cysteine is important as it is one of the major sources of sulfur.

2.5.2 Products from Proline

Various schemes have been proposed to explain the production of nitrogen-containing heterocyclic compounds, such as pyrrolidines and piperidines, from proline. Nitrogen heterocyclic compounds are often potent flavouring chemicals.

2.5.3 Strecker Aldehydes

Strecker aldehydes are produced by the Strecker degradation of the initial Schiff base (Figure 5). An α-amino carbonyl compound and

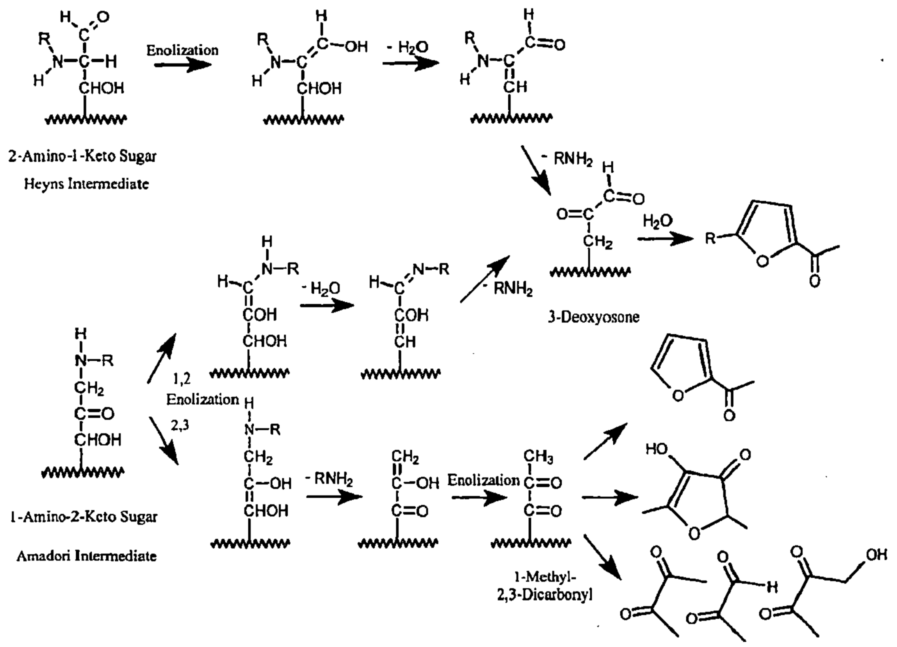

Figure 4 *Degradation of the Heyns and Amadori intermediates*
(Reprinted with permission from *Food Flavours: Biology and Chemistry*, C. Fisher and T. R. Scott, ©1997 Royal Society of Chemistry)

Strecker aldehyde are generated by rearrangement, decarboxylation and hydrolysis. Thus the Strecker degradation is the oxidative de-amination and de-carboxylation of an α-amino acid in the presence of a dicarbonyl compound. An aldehyde with one fewer carbon atoms than the original amino acid is produced. The other class of product is an α-aminoketone. These are important as they are intermediates in the formation of heterocyclic compounds such as pyrazines, oxazoles and thiazoles, which are important in flavours.

In the final stages of the reaction, brown nitrogenous polymers and copolymers form. The chemical nature of the compounds concerned is little known. Heating proteins and carbonyl-containing compounds together causes protein gels to form.[3] The proteins become covalently linked to one another. This sort of process could easily occur in toffee making. There are claims that the effect of the Maillard reaction is to reduce the availability of amino acids. As confectionery is only a minor part of the diet this is only a minor problem. If amino acids have undergone complicated reactions it is not too surprising that they are not biologically available in the finished product.

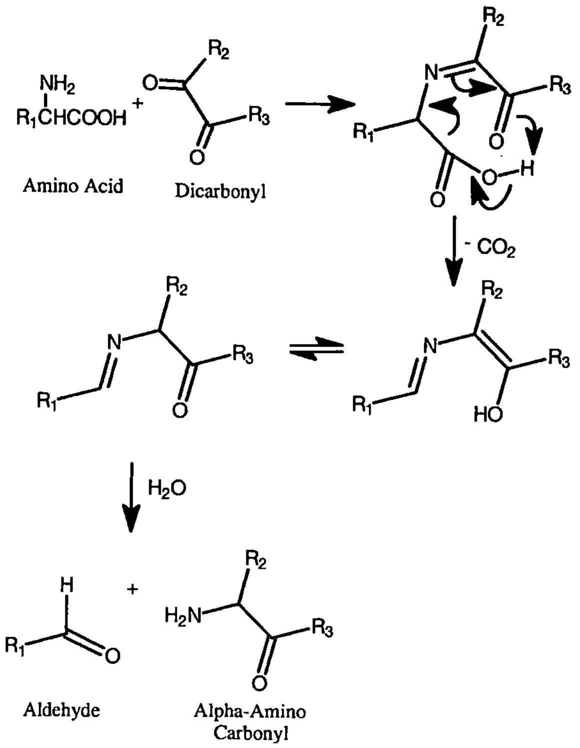
NH2
R1CHCOOH
+
O
R2
R3
O
Amino Acid
Dicarbonyl
R2
N
R3
R1
O
O
O
H
- CO2
R2
N
R3
R1
O
R2
N
R3
R1
HO
H2O
H
+
R1
O
R2
H2N
R3
O
Aldehyde
Alpha-Amino
Carbonyl

Figure 5 *Strecker degradation*
(Reprinted with permission from *Food Flavours: Biology and Chemistry*, C. Fisher and T. R. Scott, ©1997 Royal Society of Chemistry)

2.6 DENSIMETRY

The density of sugar syrups is used as a method of measuring the quantity of sugar present. Very accurate measurements of density are possible with the best equipment; confectioners often use simple hygrometers. The available data give very accurate information relating density to sugar concentration.

Some non-SI units are in use in this area. Rather than report a density the ratio of the density to that of water, *i.e.* the specific gravity, is used. This of course makes the specific gravity a ratio and hence independent of the units used. The percentage of sucrose by weight is sometimes reported in degrees Brix. The difference between reporting sucrose concentrations as weight/weight (w/w) and weight/volume (w/v) can be considerable. As an example, 50 g of sugar in 50 g of water is 50% sugar w/w, *i.e.* 50 Brix, but 50 g of sugar dissolved in water and made up to 100 mL is 50% w/v, which is approximately 42% w/w; 50 g of sugar dissolved in 100 mL of water approximates to 33.3% w/w.

The Baume scale is still used in the industry where: Baume $= M - (M/S)$. Here, M is a modulus and S is the specific gravity. In the UK, $M = 144.3$ while in the USA and parts of Europe $M = 145$.

Tables have been published relating Baume, Brix and specific gravity. As density is temperature dependent it is necessary to either bring the syrup to a fixed temperature or, as is more common in practice, to use temperature correction factors or tables. The relationship between density and concentration is slightly different for invert sugar or glucose syrups. The Brix scale is sometimes applied to products that are not sucrose syrups, such as concentrated fruit juice. Recipes are certainly in use that state "boil to x Brix". In practice these instructions mean that the material should give the same reading as a sugar syrup of that concentration. As often happens in confectionery these practices have been proved to work empirically.

2.7 REFRACTIVE INDEX

Another commonly used control measure employs the refractive index. The refractive index of a substance is the ratio of the velocity of light in a vacuum to the velocity in the substance (Figure 6). When light passes from one medium to another the beam is refracted according to the change in the refractive index. The variation in refractive index with concentration for sucrose is well known. Similar but not identical variations occur for glucose and invert sugar syrups. In practice, refractometers calibrated to measure sucrose concentration are normally used regardless of the actual sugars present. Apart from the boiling point, refractive index is the commonest control measure used in the manufacture of sugar confectionery. A refractometer is normally more expensive than a thermometer.

2.8 BUFFERS

Buffers are a convenient way of obtaining a fixed pH. Some natural materials, *e.g.* fruit juices, have considerable buffer capacity. In confectionery, buffers are used as part of fruit flavour systems and when using high methoxyl pectin. With high methoxyl pectin, gelling only takes place at high soluble solids and at acid pH. A buffer might consist of the sodium salt of a weak acid, *e.g.* boric acid and sodium borate. Because the weak acid is only feebly dissociated and the sodium salt is essentially completely dissociated, adding acid or alkali merely displaces the equilibrium, maintaining the pH. Almost all pHs can be obtained by appropriate choice of buffer.

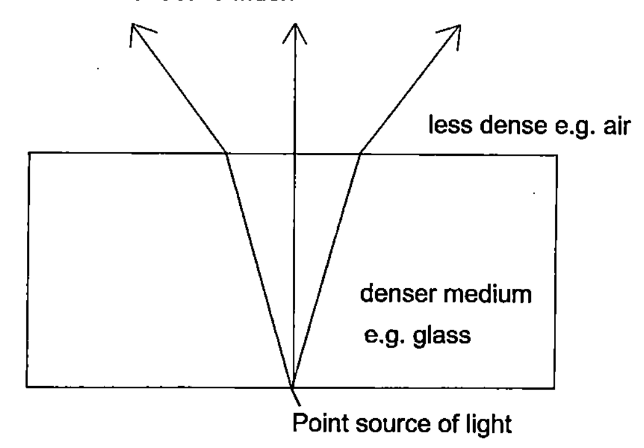

Figure 6 *A refractometer*

2.9 ANALYTICAL CHEMISTRY

The analytical chemistry that is applied to confectionery as in other products has changed enormously. High powered analytical techniques are now readily available (see also Chapter 4).

2.9.1 Water Content

The amount of water present is fundamental to the stability of confectionery products. Unsurprisingly, therefore, measuring water content is an important exercise. Various methods are used. Some oven drying moisture content determinations are still carried out. This sort of work is difficult since the moisture contents are normally low and the samples can only be dried with difficulty. In particular, there are problems in drying the product in a reasonable time without charring it. Various other methods of water content determination are in use. One such is the Karl Fischer titration.

In this system a reagent originally prepared by reacting sulfur dioxide and iodine dissolved in pyridine and methanol is used. The reaction can be represented as follows:

$$3C_5H_5N + I_2 + SO_2 + H_2O \longrightarrow 2C_5H_5NH^+I^- + C_5H_5N^+\!-\!SO_2\!-\!O^-$$

$$C_2H_5N^+\!-\!SO_2\!-\!O^- + CH_3OH \longrightarrow C_5H_5N\begin{matrix}\diagup OSO_2OCH_3\\ \diagdown H\end{matrix}$$

Initially, the sulfur dioxide is oxidised by the iodine. This can take place only in the presence of an oxygenated molecule. The product can be regarded as pyridine sulfur trioxide. In the next stage the methyl ester is formed. Thus one molecule of water is equivalent to one molecule of iodine. The original Karl Fischer reagent is prepared with excess of methanol. The methanol acts both as a solvent and as a reagent in forming the complex. This type of reagent tends to be unstable. Alternative forms of Karl Fischer regent substitute ethylene glycol monomethyl ether (methyl Cellosolve™) for methanol. Versions of the reagent without pyridine are available. The pyridine-free reagent tends to be less successful than the original form.

Although it is just possible to perform Karl Fischer titrations using the iodine colour as an indicator and ordinary burettes situated in a fume cupboard, special titration apparatus is normally used. The end-point is normally measured electrically by applying a small voltage across two platinum electrodes. In the usual form of the apparatus the sample is titrated with the Karl Fischer reagent. At the end-point free iodine appears, causing an increase in conductivity. This can be detected electronically. Modern Karl Fischer titrators aim for a high degree of automation. Some instruments have a macerator blade in the titration vessel to break up the sample. This is effective with brittle samples such as boiled sweets where the sample shatters on impact with the blade. It tends to be unsatisfactory with gum and jelly sweets, which tend to be rubbery and instead of shattering remain intact and only release their water content slowly.

2.9.1.1 Instrumental Methods. Water determinations probably tend to work well on instrumental analysis because water is radically different from other substances. Methods such as NMR and near-infrared spectroscopy are both applied to confectionery products.

NMR. Proton NMR is obviously likely to give an enormous range of signals from a typical confectionery product. An NMR instrument to analyse water in foods has to be a low-resolution instrument, whether of the original continuous form or of the later pulsed type. The aim is to discriminate between the protons in water and those in other molecules. Fortunately, this is not too difficult.

Near-infrared Spectroscopy. Near-infrared spectroscopy (NIRS) uses that part of the electromagnetic spectrum between the visible and the infrared. This region has the advantage that the instrumentation is nearest to visible instrumentation. Signals in the near-infrared come not from the fundamental vibrations of molecules but from overtones. As well as the instrumentational advantages, because the signals come from overtones the selection rules are relaxed and all possible absorbances occur. In general, NIRS measures the overtones of stretches using OH for water and NH for protein. As water gives a response different from other substances this determination often works well.

2.9.1.2 Problems with Moisture Determination. It might be thought that measuring the moisture content of foods being dried would then be easy. This is obviously a useful control measure in a factory where gums or pastilles are being stoved. However, the sweet is not homogeneous. It would be entirely possible to have a dried sweet with outside and middle solids contents of 92% and 86%, respectively. Any surface-biased technique could produce any value between 92% and 86% on a cross section of the same product.

2.9.2 Sugar Analysis

An old fashioned chemist would perform sugar analyses by a mixture of Fehling's titration before and after inversion and polarimetry. As sugars are non-volatile, it is not possible to use gas chromatography to analyse them directly. If sugars are to be analysed by gas chromatography they must first be derivatised to produce volatile derivatives. The first big improvement in direct sugar analysis came in the use of HPLC. The columns used were silica-based amino bonded phases with a mobile phase of acetonitrile and water. Polymer-based metal-loaded cation exchange columns were also used. These methods worked well for small sugars but were hampered in some ways. The detector of choice was the refractive index detector since sugars do not have a UV absorption, except at short wavelength. Short wavelength UV detection requires especially pure acetonitrile and there are many interferences. The refractive index detector precludes the use of gradient elution. HPLC was not generally able to analyse a whole range of large and small saccharides in one chromatogram. One problem when analysing the sort of mixtures of sugar and glucose syrup common in sugar confectionery was that the high molecular weight component of the glucose syrup was not eluted and had periodically to be washed from the column.

Ion chromatography has become available and is used for sugar analysis. In this system a high-performance anion exchange column is used at high pH. This separation works because neutral saccharides behave as weak acids. Table 1 shows some pK_a values.

Very conveniently, the technique can also handle the sugar alcohols such as sorbitol. Detection is by pulsed amperometric detection. This

Table 1 *pK_a values of some sugars*

Sugar	*Fructose*	*Mannose*	*Xylose*	*Dextrose*	*Galactose*	*Sorbitol*
pK_a	12.03	12.08	12.15	12.28	12.39	13.6

system works by detecting the electrical current generated by oxidation of the carbohydrate at a gold electrode. The oxidation products poison the surface of the electrode, necessitating cleaning between measurements. Cleaning is carried out by raising the potential to oxidise the gold surface. This causes the oxidation product to desorb. The potential is then lowered, which reduces the electrode back to gold. Thus the sequence of pulsed amperometric detection measures the current at the first potential then applies a more positive potential to oxidise and clean the electrode followed by another potential to reduce the electrode back to gold, ready for the next detection cycle. In operation, the three potentials are applied for a fixed duration. There is a charging current when changing potentials. The oxidation current is distinguished from the charging current by measuring it after the charging current has decayed. Integration of the of the cell current over time is used to obtain the carbohydrate oxidation current. As the integration of current over time gives charge, the value obtained is in coulombs.

An important question is how this system can work with sugar alcohols and non-reducing sugars. The oxidation is catalysed by the electrode surface, which means that the response is dependent on the electrode potential of the catalytic state rather than the redox potential.

As the mobile phase in this system is normally a sodium hydroxide solution there is no need to handle or dispose of organic solvents. This is a particular bonus to some smaller sites that are not set up to use organic solvents.

An important issue for any laboratory analysing products rather than raw materials is the extraction of the material of interest from the product for analysis. Some sugar confectionery such as boiled sweets can be dissolved directly with little preparation. Other materials like toffees require considerable extraction and clean up. If a toffee is to be analysed for sugars the sugars have to be separated from the proteins and fat present. This is made more difficult by the fact that the system is by design a stable emulsion. Most methods of sugar analysis require a clean aqueous extract to work on. One problem of working with HPLC columns is that minor components can accumulate on, and thereby deactivate, the column. Ion chromatography columns, however, are relatively easy to clean because sodium hydroxide at high pH can be used.

An analytical method for a butterscotch would run as follows: dilute the sample 1:2000 and pass through a 0.2-micron filter. The dextrose, fructose, maltose and maltotriose can then be measured directly. In contrast, the same analysis would probably require two chromatograms performed with different mobile phases using the amino-column HPLC method.

2.10 EMULSIONS

An emulsion is a dispersed system of two immiscible phases. Emulsions are present in several food systems. In general, the disperse phase in an emulsion is normally in globules 0.1–10 microns in diameter. Emulsions are commonly classed as either oil in water (O/W) or water in oil (W/O). In sugar confectionery, O/W emulsions are most usually encountered, or perhaps more accurately, oil in sugar syrup. One of the most important properties of an emulsion is its stability, normally referred to as its emulsion stability. Emulsions normally break by one of three processes: creaming (or sedimentation), flocculation or droplet coalescence. Creaming and sedimentation originate in density differences between the two phases. Emulsions often break by a mixture of the processes. The time it takes for an emulsion to break can vary from seconds to years. Emulsions are not normally inherently stable since they are not a thermodynamic state of matter. A stable emulsion normally needs some material to make the emulsion stable. Food law complicates this issue since various substances are listed as emulsifiers and stabilisers. Unfortunately, some natural substances that are extremely effective as emulsifiers in practice are not emulsifiers in law. An examination of those materials that do stabilise emulsions allows them to be classified as follows:

(i) surfactants
(ii) natural products
(iii) finely divided solids

Some substances fall into more than one category. In a practical emulsion system the emulsifier should facilitate making the emulsion while stabilising it after formation. Some properties have opposite effects in these two areas. A high viscosity makes it harder to form an emulsion but obviously tends to stabilise the emulsion when formed.

2.11 THE CHEMISTRY OF OILS AND FATS

Fats are, chemically, triglycerides and can be regarded as the esters produced by the reaction of fatty acids with the trihydric alcohol glycerol. In practice oils and fats are the product of biosynthesis. Some sugar confectionery contains oils or fats while other products, *e.g.* boiled sweets, are essentially fat free. The traditional fat used in flour confectionery is milk fat, in the form of butter, cream, whole milk powder or condensed milk. Milk fat can only be altered by fractionating it. While this is perfectly possible, there has to be sufficient commercial and technical benefits to make it worthwhile. One problem with fractionation operations is that both the desirable and the undesirable fraction have to be used.

While vegetable fats were used originally as a cheaper substitute for milk fat the ability to specify the properties of vegetable fat has considerable advantages. This ability arises because of the science and technology available to the fat processing industry. Some vegetable fats used in foods are not tailor-made but are simply a vegetable fat of known origin and treatment. The commonest example is hydrogenated palm kernel oil (HPKO), which is often used in foods.

Some fats go into confectionery as a component of other ingredients. The common example is nuts, which contain fats, often of types such as lauric or unsaturated fats. These fats are sometimes the origin of spoilage problems.

2.11.1 Classifications of Fatty Acids

Fatty acids consist of a hydrocarbon chain with a carboxylic acid at one end. They can be classified on the basis of the length of the hydrocarbon chain and whether there are any double bonds. Trivial names of fatty acids such as butyric, lauric, oleic or palmitic are in common use in the food industry. A form of short hand is used to refer to triglycerides, where POS is palmitic, oleic, stearic. If the chain length is the same an unsaturated fat will always have a lower melting point. Another classification of fats is in terms of the degree of unsaturation of the fatty acids. Saturated fats are those without any double bonds. Many animal fats are saturated but some vegetable fats, *e.g.* coconut oil, are saturated. Monounsaturated fats include oils like olive oil but also some partially hydrogenated fats. Polyunsaturated fats have many double bonds and include sunflower oil. They are not normally used in foods as they are too unstable. The only exceptions to this would be if a polyunsaturated oil, *e.g.* sunflower oil, was used for marketing reasons. Other occasions have occurred when sunflower oil has been used by those unaware of its chemistry. Traces of polyunsaturated oils do go into foods as components of ingredients, *e.g.* nuts.

2.11.2 The Hydrogenation of Fats and Oils

To provide the right properties is often necessary to reduce the degree of unsaturation of a particular fat or oil. This is achieved by hydrogenating the oil over a catalyst, usually nickel. Hydrogenation can be either complete, yielding a saturated fat, or partial, yielding a partially hydrogenated or hardened fat. Partial hydrogenation tends to produce fats with trans double bonds. A double bond is physically flat and does not permit rotation. Thus the chemical groups are fixed in their relation (Figure 7). A cis double bond is one where the hydrogen atoms are both on the same side. In contrast, a trans double bond has them on the opposite side. Most naturally occurring oils and fats have cis double bonds; however, some trans double bonds are found in milk fat and some marine oils.

2.11.3 Fat Specifications

Apart from specifications as to origin, *e.g.* palm kernel oil, fats are normally supplied on the basis of established parameters. One of these is the iodine value. This reflects the tendency of iodine to react with double bonds. Thus, the higher the iodine value the more saturated the fat is. An iodine value of 86 would approximate to one double bond per chain, while an iodine value of 172 approximates to two double bonds per chain. Another parameter is the peroxide value. This attempts to measure the susceptibility of the fat or oil to free radical oxidation. The test is applied on a freshly produced oil and measures the hydroperoxides present. These hydroperoxides are the first stage of the oxidation process. Obviously, this test would not give reliable results if applied on a stale sample.

2.11.4 Deterioration of Fats

Fats deteriorate in two ways. One is normally a chemical process the other is normally enzymatic. They are oxidative rancidity and lipolytic rancidity. In the former, oxygen (normally in the form of a free radical) adds across double bonds. As this is a zero activation energy process it is

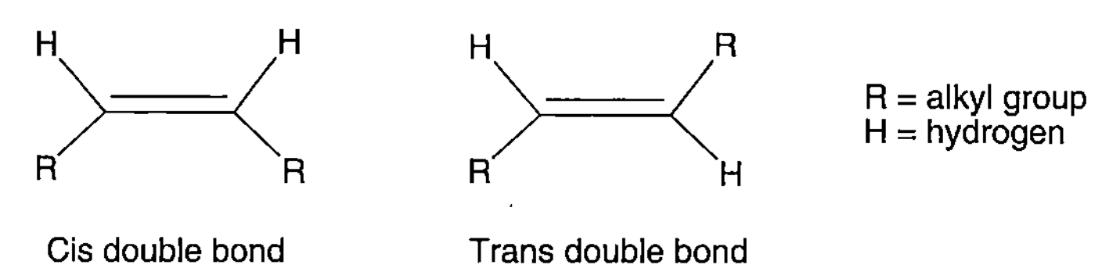

Figure 7 *Stereochemistry of cis and trans double bonds*

not inhibited by reducing the temperature. Oxidative rancidity is prevented by using saturated fats. The end-products of this process can be unpleasant tasting and smelling. Oxidative rancidity tends to appear suddenly and then progress rapidly. Lipolytic rancidity is normally enzymatic. The enzymes responsible usually come from bacteria or moulds. The effect of lipolytic rancidity is that the level of free fatty acid rises. The effect of this on the product depends very much on the nature of the free fatty acid liberated. Low levels of free butyric acid from milk fat tend to enhance a toffee by giving it a more buttery flavour. In contrast, lipolysis of a lauric fat such as HPKO affords free lauric acid. Lauric acid is an ingredient of and tastes of soap. This effect is very unpleasant.

2.12 WATER MIGRATION

A common problem in baked products occurs where the baked material is in contact with another material with a higher water activity. Baked products tend to have a low water activity and will soften if they pick up water.

As an example, Cauvain and Young give figures for the water activity of the components of savoury pies such as pork pies as pastry (0.24), jelly (0.99), and filling (0.98).[4] The jelly and the filling clearly have nearly matched water activities so migration is unlikely to be a problem.

There are two ways of preventing this problem, the first is to modify the high water activity material or provide a barrier layer; the second is to decrease the water activity of the material that is in contact. Examples of this sort of problem are ice cream in contact with a wafer, caramel in contact with a wafer and the filling and the pastry of an apple pie.

2.12.1 Barrier Methods

Any barrier used has to be edible of course. An obvious possibility is to place a fatty barrier in the product. This barrier has ultimately to be palatable and ideally to add to the attractiveness of the product. The barrier has to be carefully applied without any obvious gaps. It does not have to provide a hermetic seal, merely retard migration sufficiently to prevent the product softening before it is spoiled for some other reason, *e.g.* the meat component of a meat pie will spoil quite quickly.

A successful example is ice cream products that are sold from the freezer and consist of a wafer cone filled with ice cream. If the ice cream was left in contact with the cone the cone would become soft. This is solved by adding a chocolate-flavoured coating to the inside of the cone.

The coating prevents contact between the cone and the ice cream, thereby preventing the migration of water. Such a chocolate flavoured coating meets the required criteria, it is edible, effective and it adds to the attractiveness of the product.

2.12.2 Matching the Water Activity

In practice, to maintain the right texture of a bakery product the water activity has to remain low. Therefore, the only way of matching the water activity of the two components is to lower the water activity of the other component.

2.13 THE SCIENCE OF PROTEINS

2.13.1 History

Proteins have been studied for a long time. Beccari published an account of his experiments to isolate gluten in 1747! In 1805 Einhof discovered that a fraction of wheat gluten was soluble, while in 1858 Denis showed that many proteins of both plant and animal origin were soluble in saline solutions. In 1859 Ritthausen started to prepare highly purified proteins, only to be criticised by Weyl for using alkali to extract the proteins. Weyl in his work used the Denis method of extraction with neutral salts.

Work on the solubility was continued by Osborne to produce the classifications of vegetable proteins that are still used. Studies of the solubility of proteins remains important both in characterising and purifying proteins.

The word protein was invented by Mulder, who believed that there was a protein radical that was common to all proteins.

The history of protein chemistry can be summarised as improved purification techniques combined with new methods of protein characterisation allied to theoretical interpretation of the results obtained. Improved computational techniques have also helped considerably.

2.13.2 Classification of Cereal Proteins

It is possible to classify proteins on several different bases. One basis is on the cereal from which they come and where in the seed they are found, another is the Osborne classification system, which is based on the solubility of the protein. Proteins can be further classified in chemical terms such as molecular weight and the presence or absence of sulfur.

Cereal proteins when classified by the Osborne sequential extraction method yield four different classes: albumins, which are water soluble, globulins, which are soluble in salt solutions, prolamins, which are soluble in alcohol–water mixtures, and glutelins, which are soluble in dilute acid or alkali. Chen and Bushuk added a fifth fraction by dividing the glutelin into two fractions, one soluble in dilute (0.05 M) acetic acid and the other insoluble in this reagent.[5]

Of course these classes do not constitute a single protein but mixtures of proteins having the same solubility characteristics. Improved methods of fractionation reveal how many different components are present in each fractionation.

Biochemical and genetic researchers divide the proteins into three groups: S-poor, *i.e.* sulfur poor, S-rich, *i.e.* sulfur rich, and HMW-prolamins, *i.e.* high molecular weight prolamins.

The Osborne classification, which dates from 1907, was updated at a symposium on gluten in 1996. Gianibelli *et al.* point out that at least 1300 peptides can be obtained from wheat endosperm proteins after disulfide bond rupture using two-dimensional fractionation.[6]

Although the Osborne classification has been a major milestone in cereal chemistry, the gliadins and glutenins (the two classes of protein found in gluten) overlap in their solubilities. This led to arguments over whether particular fractions are low molecular weight glutenins or high molecular weigh gliadins. Gliadins are single polypeptide chains, *i.e.* monomeric proteins, while glutenins consist of many chains of polypeptides held together by disulfide bonds.

2.13.3 Glutenins

Glutenins are among the largest protein molecules in nature,[7] with molecular weights of 20×10^6 daltons being recorded by gel filtration.

Glutenins are heterogeneous mixtures of polymers formed from polypeptides linked by disulfide bridges. These are classified into four groups on the basis of their mobility in sodium dodecyl sulfate polyacrylamide gel electrophoresis (SDS-PAGE) after the S–S bonds have been chemically reduced. The four groups are called A, B, C and D after the four regions of electrophoretic mobility. The A group, which has a molecular weight of between 80 and 120 kDa, corresponds to high molecular weight glutenin subunits (HMW-GS), while the B (42–51 kDa) and C groups (30–40 kDa) are low molecular weight glutenin subunits (LMW-GS) and are slightly related to γ- and ω-gliadins.[6]

Various new methods such as capillary electrophoresis and reversed-phase HPLC have been used to characterise glutenin subunits. These techniques, in addition to high resolution, can be used to produce quantitative results and are capable of automation.

Capillary electrophoresis separates by differences in charge while reversed-phase HPLC separates on the basis of hydrophobicity. It turns out that these proteins have wide differences in hydrophobicity.

2.13.3.1 Glutenin Extraction. One problem that new technology has not altered is the need to extract glutenins prior to further work. Any extraction has two problems: first the need to completely extract the material of interest and, secondly, to not extract anything else. The methods used have included precipitation with 70% ethanol, pH fractionation, an initial solubilising step treating the gluten with 0.05 M acetic acid, and pH precipitation with ion exchange chromatography. As a result of these investigations it emerged that the glutenin extracts obtained were significantly different in composition and some contained large amounts of previously unextracted gliadins! Treating the extracts subsequently with 70% ethanol removed the gliadins.

Other workers used 0.1 M acetic acid for gluten separation then changed to dilute hydrochloric acid followed by neutralisation with sodium hydroxide.[8] Byers *et al.* used 50% propan-1-ol in preference to 70% ethanol.[9] Methods based on extraction with sodium dodecyl sulfate (SDS) have been developed by Danno[10] as well as Graveland *et al.*[11,12] Sonication of the SDS extract was introduced by Singh *et al.*[13,14] Burnouf *et al.* introduced the use of dimethyl sulfoxide (DMSO) to remove monomeric proteins and a few small gliadins.[15]

In 1991 two quick one-step methods were developed to extract glutenins. The Singh method[16] gives complete extraction while the Gupta method[17] provides purer glutenin but gives a less complete extraction.

In 1984 Bietz was the first to use size-exclusion HPLC on glutenin extracts that had not been treated with a reducing agent.[18] Further improvements were made by using sonication and suitable extractants by Singh *et al.*[13], Batey *et al.*[19] and Gupta and Shepherd.[20] Size-exclusion HPLC followed by ion exchange chromatography was used by Lew *et al.* to both isolate and purify glutenin.[21] Osborne's fractionation scheme was combined with SE-HPLC by Huebner and Bietz.[22] A new fractionation method based on differences in solubility in 50% and 70% propan-1-ol has been developed by Fu and Sapiratein.[23] The above list is not exhaustive and work in improving this area continues.

2.13.3.2 The Importance of Studying Gluten Fractions. What use, it could be argued, are exercises in fractionating gluten proteins? It is from this research that the importance of the different fractions in bakery performance can be deduced. While this is not yet a source of useful information to the baker the ability to determine the presence or absence of given fractions is useful to the plant breeder. If laboratory-scale quantities of wheat can be tested, breeding new wheat varieties for particular purposes becomes easier. Varieties that are likely to have poor bread-making properties, for example, can be abandoned early in the process.

2.13.3.3 The Importance of High Molecular Weight Glutenin Subunits. While these components are present in small quantities they are very important in terms of gluten elasticity, which is desirable for bread making. It appears that these components encourage the formation of larger glutenin polymers.

2.13.3.4 The Importance of Low Molecular Weight Glutenin Subunits. The study of these subunits is less advanced than that of the higher molecular weight materials because the lower molecular weight glutenins have been harder to purify. Some glutenin subunits act as chain extenders, giving stronger doughs, while others act as chain terminators giving weaker doughs.

2.13.4 The Importance of Gliadins

Gliadins, unlike glutenins, contribute to the viscosity and the extensibility of the gluten and not its strength.

2.13.5 The Wheat Seed

The wheat kernel or caryopsis is divided into three anatomical regions: the bran, the embryo, and the endosperm. Bran consists of the pericarp, the testa, the nucellar layer and the aleurone layer. The components of bran are fibre (comprising hemicelluloses, β-glucans, cellulose and glucofructans,), minerals, enzymes, vitamins and globulin storage proteins. The embryo contains lipids, enzymes (lipases and lipoxygenase), vitamins and globulin storage proteins. The largest proportion of the wheat seed is the endosperm, which contains the nutrients necessary for germination.

There is a composition gradient across the grain, with the percentage of both starch and protein rising. There is also a gradient in protein quality, with the protein nearest the centre having the best baking quality. Apart from the desire for whiter products this quality gradient is the reason for the continued use of patent flour.

2.13.6 Enzymes

The enzymes in wheat, and hence in flour, that often cause problems in the bakery are present in the seed to make nutrient available to the seed. Similarly, this is why sprouted wheat causes problems if it is allowed to get into flour. Thus, the α-amylase is low in mature wheat grains but rises rapidly on germination. In bread, a low, but not too low, level of α-amylase is desirable since it produces sugars to feed the yeast and opens up the structure. Deliberate additions of malt flour were once common, but are now rarely made, to increase the amylase level.

About 80% of the β-amylase present in wheat flour is associated with glutenins. Although acid proteases and peptidases are present in the wheat seed they do not normally cause problems in the bakery.

Both lipases and lipoxygenases are present in the bran and the germ. Phytases are nutritionally important as they liberate the phosphorus, of which approximately 70% is in the kernel bound to phytin. Phytin blocks the intestinal absorption of both iron and calcium. Phytase is also present in yeast, which is why leavened bread is nutritionally superior to unleavened bread. There have been concerns about the incidence of rickets among those of South Asian origin who eat chapattis, live in the UK, and have a tendency to keep their skin covered up from the sun.

Both peroxidase and catalase are present, with peroxidase causing the oxidative crosslinking of pentosans in the dough. Similarly, glutathione is oxidised by glutathione dehydrogenase during the process of making dough; this is a rapid reaction and promotes disulfide interchange in sulfur-containing proteins.

Enzymic actions that depolymerise proteins such as the high molecular weight glutenins will cause the viscosity of the dough to drop. This outcome is predictable on the basis of polymer science.

A class of enzymes that sometimes causes problems is the polyphenoloxidases, which are present in the outer layers of the kernel. These enzymes can cause dark spots to appear in pastry offcuts that are being recycled, *e.g.* to make pie cases. The enzymes act on any minute bran particles that are present.

2.13.7 Wheat and its Proteins

The proteins that are of most interest in baking are the proteins from wheat. The wheat that is normally used is *Triticum aestivum* not *Triticum durum* which is used to make pasta. The distinguishing property of wheat proteins is that some of them can develop into the viscoelastic mass known as gluten. The only other grain that has proteins with similar properties is rye, but the protein development is less effective.

Given that some of the protein, particularly that in wheat bran, does not form gluten it is obvious that the quality of the protein is as important as the quantity. Unfortunately, measuring the quantity of protein is easier and less controversial.

The quantity of protein present is normally measured by the Kjeldahl method. As wheat endosperm protein is around 17.5% nitrogen a factor of 5.7 is normally used to convert Kjeldahl nitrogen measurements into protein. Tkachuk suggested that 5.7 should be used for whole wheat but 5.6 should be used for flour.[24]

A simple test that gives more information about protein quality is the wet gluten test. Here, a flour and water dough is washed to remove the starch, leaving the gluten. The properties of the resulting wet gluten can be assessed by stretching it. Experienced testers can rapidly form an opinion on the suitability of the gluten sample for bread making. A problem that can occur is a sample that otherwise measures well but has gluten that will not stretch without breaking. This might be due to overheating of the grain when it was dried on the farm. A biscuit flour would probably fail this test, but having good gluten is not a requirement for biscuit flours. Attempts have been made to carry out the wet gluten test on a quantitative basis. The results are unreliable because of the varying efficiency of the starch washing and the weight of water absorbed by gluten.

2.13.8 The Composition of Gluten

Gluten is a mixture of proteins that can be classified as glutenins and gliadins. Extensibility is provided by glutenins while the gliadins contribute elasticity and cohesiveness.

2.13.8.1 Dried Gluten. This material is sold as vital wheat gluten, it is produced by a scaled up version of the process for producing wet gluten for flour testing. A flour and water dough is made and then washed to remove the starch, the soluble proteins and the pentosan-based gum fraction. This latter fraction is an example of a non-starch carbohydrate.

The starch-reduced dough can be used to produce starch-reduced breads and rolls for special dietary needs. Vital wheat gluten is the product produced by drying the water washed gluten. Although it is carefully dried, the dried gluten does not have all the properties of fresh gluten. Dried gluten is used to increase the protein content of flour. Its use became popular in the UK when the CAP made the use of non-EU high protein wheats expensive – a combination of EU wheat flour and dried gluten with or without some non-EU wheat flour was found to give satisfactory results, particularly in rapid bread processes such as the Chorleywood bread process (CBP). Dried gluten is less successful in the old fashioned long processes.

The combination of dried gluten and some English wheat flour allows the production of all-EU bread flours that will work in CBP plants. Thus, two lots of EU wheat have been used instead of some non-EU wheat. Of course, the by-product of making dried gluten is starch, for which some profitable use is sought. At the time of writing there is a prospect that such by-products might be fermented to produce ethanol that can be used as a petrol substitute in spark ignition engines.

It appears that dried gluten produced from English non-bread making flour is just as good as gluten produced from the best quality Canadian flour. Unsurprisingly, dried gluten is normally made from the cheapest possible source.

There is another type of gluten available and that is produced in France by air classifying wheat flour. Air classification works by separating materials in an air stream on the basis of density. This air classified gluten has not been wetted and dried and behaves more like the gluten found in flour.

2.13.9 The Utility of Research on Flour Proteins

A question that occurs to those involved in the baking industry is what use has this research on proteins been regarding producing baked products? The research has in general aimed at identifying those proteins that are most important in making bread. This is because of the economic importance of bread and the priorities of the funding providers.

To date, the research has been used to investigate different cultivars of wheat so that the benefit to the baking industry is likely to be in enhanced supplies of wheat in both quantity and quality.

2.14 THE SCIENCE OF STARCH

Starch is a major component of almost all baked products. It most commonly is incorporated into products in the form of wheat flour but various forms of nearly pure starch such as corn flour (maize starch), wheat starch and potato starch are occasionally used.

While wheat protein is important in products such as bread, starch is important in any product made from flour. The mess that occurs if bread is made from flour with too much amylase present shows what happens if the starch is broken up by the amylase.

Starch is the major energy store of plants; chemically it is a polymer of glucose and occurs in two separate forms, amylose and amylopectin. The ratio of the two types depends on the plant that the starch has come from; typically starch is 20–30% amylose and 70–80% amylopectin but there are amylomaizes with more than 50% amylose while waxy maize produces almost pure amylopectin with less than 3% amylose.

The ratio of amylose to amylopectin produced by a plant is under genetic control. Waxy maize is a natural mutation of maize that happens to produce almost pure amylopectin.

If starch with different properties is required then either a different plant source must be used, *e.g.* waxy maize instead of potato, or a way of altering the starch produced by a plant must be found.

The sort of starch produced by a plant can be altered by plant breeding to produce plants with the right genetic make up to produce the desired starch. Sometimes the desired genetic characteristics can only be found in a wild variety of the crop.

While it is possible to use genetic engineering techniques to manipulate the sort of starch produced, at the time of writing the use of such starch in foods is illegal in Europe. The starch from genetically modified plants can, however, be used in industrial products such as adhesives.

Starch occurs in granules in plants. These granules can be shown to have an organised structure both by X-ray crystallography and because they are birefringent. The birefringent properties of starch can easily be demonstrated by using a microscope with crossed polarisers. The first polariser polarises the light illuminating the sample while the second polariser is set at 90°. This gives, of course, a completely black field. If any birefringent material such as starch is placed under the microscope then the individual granules appear like stars against a night sky. This effect occurs because the starch is optically active, *i.e.* the molecule has the ability to rotate the plane of polarised light. This effect only occurs with molecules that have an asymmetric centre, *i.e.* they can not be superimposed on their own mirror image. Glucose is optically active so it is to be expected that any molecule produced from it would also be optically active. In fact, most natural products are optically active.

2.14.1 Gelatinisation

One of the important properties of starch is that it undergoes gelatinisation. Starch is insoluble in cold water and intact starch granules do not absorb cold water. However, it is possible to prepare a dispersion of starch in cold water. If such a dispersion is heated there will be a temperature at which the granules start to swell – at this point the birefringence will disappear and the viscosity will rise sharply. The weaker hydrogen bonds between molecules break as the granule swells in taking up water, causing the increase in viscosity. Inside the granule the ordered structure disappears, which can be detected by the loss of birefringence and by other physical methods.

The temperature at which this happens will depend on the origin of the starch, with starch from different plants having different characteristics. Indeed, the gelatinisation temperature is a way of determining the origin of a particular sample of starch. Changes in starch on gelatinisation can be followed by several techniques although the optical activity and the viscosity are the most common. Different techniques measure a different process and hence tend to give a different point as that at which starch is gelatinised.

The gelatinisation temperature of starch is affected by the presence of sugars, fats and salts. In any practical baked product some or all of these are likely to be present, so the gelatinisation temperature will not be that observed for the equivalent pure starch in distilled water. What is important is that the starch must be gelatinised or the product will collapse.

2.14.2 Retrogradation

Retrogradation is another important property of starch. It is generally accepted that retrogradation is involved in the staling of baked products such as bread. In particular it appears that retrogradation is the recrystallisation of the amylopectin present. Notably, retrogradation is still a subject of research. The application of techniques such as ^{13}C NMR allows insights that older techniques do not provide.

This recrystallisation releases water. This is probably why, although deep freezing prevents starch retrogradation, refrigeration at temperatures above zero causes bread to stale faster than storage at ambient temperature.

2.14.3 Starch Molecules

Given the size of the starch molecule, starch was classified as a complex carbohydrate, which chemically it is. Nutritionally, starch, particularly potato starch, is broken down into glucose fairly quickly. If the gluten molecule is regarded as a giant construction set the body has the key to break down the links between the individual molecules. However, while some starch breaks down fairly quickly other starch does not. The starch that does not is classified as resistant starch. Excluding chemically modified starch, resistant starch can be classified as: physically inaccessible, *e.g.* in whole grains of wheat, ungelatinised starch, *e.g.* raw starch, and thermally stable retrograded starch, *e.g.* as found in bread, particularly in stale bread. Thermally stable retrograded starch is mainly amylose.

The non-break down of physically inaccessible starch explains why wheat is ground into flour in the first place. This may also explain the claims that modern flour with its very small particle size is less healthy than the sort of flour produced by ancient wind and water mills. The resistant nature of ungelatinised starch also explains why starch-based foods were baked in the first place.

Resistant starch will serve as primary source of substrate for colonic microflora and may have important physiological benefits. On this basis resistant starch can be classified as a dietary fibre. The Association of Official Agricultural Chemists (AOAC) method of determining dietary fibre will measure some resistant starch as dietary fibre.

Chemically, both varieties of starch are polymers of glucose with the α-D-glucose units in the $^4C_1$ conformation. The glucose units are linked -(1→4)- in both amylose and amylopectin but in amylopectin roughly one residue in twenty is linked -(1→6)-, which forms branch points. The proportion of branch points varies, depending on the source of the amylopectin.

2.14.4 A Comparison of the Structure of Amylose and Amylopectin

Amylose has a lower molecular weight than amylopectin but forms linear chains while amylopectin has a higher molecular weight but forms more compact molecules. While both molecules have a structure mainly based on α-(1→4)-D-glucose units, the amylopectin structure is branched at the α-(1→6)-D-glucose units.

α-(1→4) links allow relatively free rotation about the phi and psi torsions of hydrogen bonds between the O3′ and O2 oxygen atoms of sequential residues, which tends to encourage the molecule to adopt a helical conformation. The helical structures can present contiguous hydrophobic structures as a consequence of their relative stiffness.

2.14.4.1 Amylose. Amylose molecules are composed of single chains, mainly but not exclusively unbranched, consisting of 500–200 000 α-(1→4)-D-glucose units. The molecular weight depends on the plant from which the amylose has come. Although a small quantity of α-(1→6) branches and linked phosphate groups may be found these appear to be too few to have any effect.

2.14.4.2 Amylopectin. Amylopectin can be regarded as the α-(1→4)-D-glucose structure of amylose with non-random α-(1→6) branches. Branching enzymes govern this branching, leaving each chain with up to 30 glucose residues. While only 5% of the glucose residues form branch points there are some one or two million or so glucose residues in an amylopectin molecule.

The outer chains, which are unbranched, are called A chains while the inner branched chains are called B chains. The single reducing group is contained in a unique chain called the C chain (Figure 8a).

A typical amylopectin molecule will contain slightly more outer, *i.e.* A, than inner, *i.e.* B, chains.

The Semicrystalline Structure of Amylopectin. Amylopectin crystallises according to a cluster molecule,[25] as shown in Figure 8b. Within this structure there are crystalline and amorphous regions. An

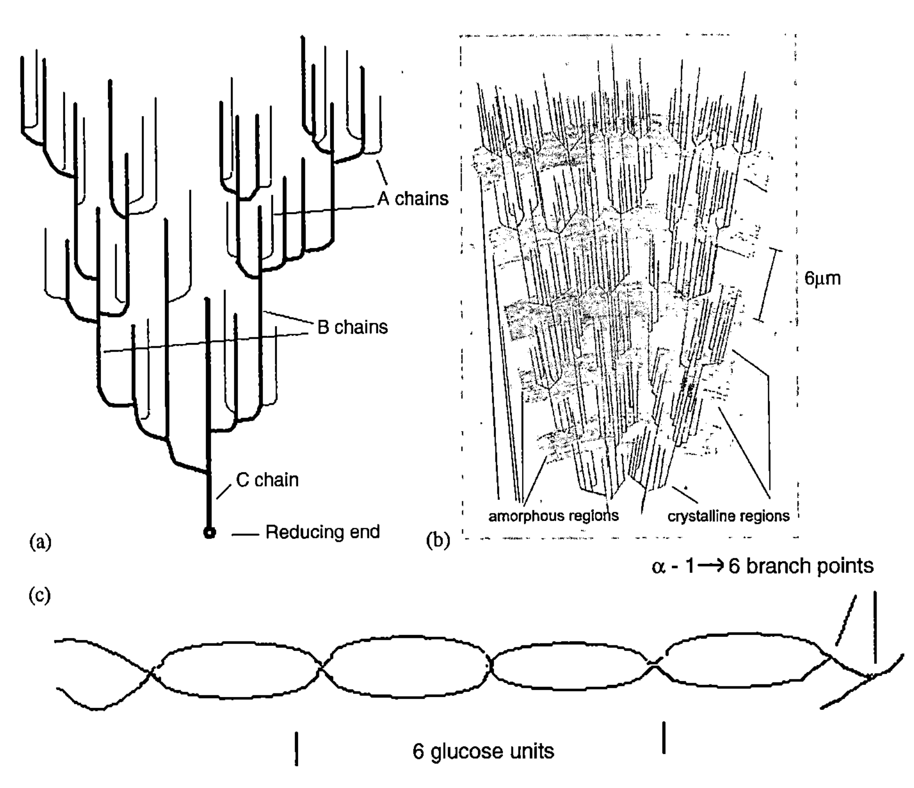

Figure 8 *The structure of amylopectin (a) organisation of the molecule; (b) crystalline and amorphous regions; (c) possible double helix structure*

amylopectin molecule has a hydrodynamic radius of around 21–75 nm. In Figure 8; (a) shows how the molecule is organised; (b) shows the arrangement of the crystalline and amorphous regions; (c) shows a possible double helix structure taken up by neighbouring chains, which is the source of the high degree of crystallinity in the granule. There is some dispute over the form of the crystalline structure but the most likely candidate is left-handed helices with six residues per turn.

2.14.5 Modified Starches

Starches with modified properties can be obtained in several ways. One way is to use a source with starch that has different properties. Examples of this are waxy maize, which gives almost pure amylopectin, and tapioca. These are still starch in food law.

Another altered starch is damaged starch, which is purely mechanically modified starch. This is starch whose granules have been damaged in the milling process. The important property of damaged starch is that, unlike undamaged starch, it absorbs water in the cold.

As a comparison, in a dough with an excess of water, the water absorption of protein, undamaged starch, and damaged starch, as measured by weight and comparing the weight of water absorbed to the original weight, is protein 200%, undamaged starch 33% and damaged starch 100%.

Damaged starch is also more readily degraded by enzymes. The ability to absorb water is the most important property of damaged starch since it adds to the ability of flour to absorb water. If all the starch in a flour is undamaged then only the protein will absorb water.

Damaged starch is desirable in British bread because it gives two advantages, the increased water absorption increases the yield while the damaged starch is more readily attacked by enzymes, thus producing sugars to feed the yeast. High levels of starch damage can only be produced in hard wheat flour, which is one of the reasons for the continued use of hard wheat flour.

It should be appreciated that a high level of starch damage is not essential in bread. French bread is made from soft wheat flour with a low starch damage. Starch damage is generally undesirable in biscuits. In biscuits the product is cooked to a very low moisture content so binding in water is undesirable.

The desirability of damaged starch varies between products, in some products damaged starch is desirable while in others it is undesirable. Even in bread too high a starch damage is undesirable as product quality suffers.

It is not particularly easy to measure the degree of starch damage present. The usual method involves treating the flour with α-amylase, which can only attack the damaged starch. The procedure requires an α-amylase preparation that has to be standardised. Alternatively, an estimate can be made by optical microscopy or by calculating from the water absorption of the flour and its protein content, assuming that the water absorption that exceeds that to be expected from the protein alone is due to the damaged starch.

Another modified starch is pregelatinised starch, *i.e.* starch that has been gelatinised and then spray dried. The resulting product gives a starch that will absorb water and form a gel with cold water. Pregelatinised starch is normally used in instant puddings and similar products.

All of the above starches still qualify as starch under food law; there are starches that have been chemically modified and hence qualify for the term "modified starch" on the label.

2.15 NUTRITION

If this work had been written a few years ago this subject would probably not have been included. The food industry is now under considerable pressure over the nutritional properties of its products, so that those working in the industry have more need of nutritional knowledge than before.

Some parts of the bakery industry are in a different situation to others regarding nutritional pressure. The parts of the industry that make bread are largely in the clear. The composition of bread is controlled anyway and bread in moderation is regarded as a healthy staple food. There are those who claim that the general population would be healthier if they ate wholemeal rather than white bread. Well, the industry makes wholemeal bread and the customers are free to buy it. If they choose not to it has to be their fault. The only nutritional pressure regarding the composition of bread has been over the amount of salt added, which the industry has agreed to lower.

The baking industry has even developed soft grain breads and other multigrain breads to supply consumers' nutritional needs.

Producers of cakes, biscuits and pastries are in a less easy position. While bread is a staple food containing complex carbohydrates and is essentially sugar free and low in fat, cakes, biscuits and pastries contain quantities of sugar and fat. They are also regarded as non-essential foods.

For most of human existence the major nutritional problem has been getting enough to eat all year round. In the western world the major problem is now obesity! The transitions to the modern situation in the western world have occurred since World War II. During the 1930s a confectionery manufacturer ran advertisements proclaiming the large number of calories per penny in its new product. No similar manufacturer would do so today. In the 1950s another confectionery manufacturer ran posters showing a steam locomotive crew sharing a bar of chocolate at the end of the journey. The fireman had probably shovelled seven or eight tonnes of coal into the firebox and needed some nourishment. The fireman's job and many similar heavy manual jobs have long since disappeared. Sedentary occupations lead to different nutritional needs.

2.15.1 Nutritional Needs

A major nutritional need is energy. The amount needed varies from individual to individual and depends on the amount of energy expended. This energy ought to be measured in the SI unit of joules but the custom and practice is to use the calorie. Properly, the unit used ought to be the kilocalorie, when calories are used in this field without a prefix the unit being used is the so-called large calorie which was written "Calorie", *i.e.* with a capital C, where 1 Calorie = 1 kilocalorie. It is redundant to inquire which of the three definitions of calorie is used as nutritional data are variable. Obviously, the energy content of foods is inherently dependent on the variability of the product produced by variability in growing conditions. However, there are other problems with energy contents. The energy content of a food is normally measured by using a bomb calorimeter. In this instrument a weighed quantity of the food under study is ignited in a sealed steel vessel with a pure oxygen atmosphere. The temperature rise is measured and the amount of energy produced is calculated from the same temperature rise produced in the same calorimeter by a known amount of energy. This is exactly the same procedure that is used to determine the energy content of fuels like coal. One problem with foods is that some materials are incompletely metabolised, thus the bomb calorimeter value is not that achieved in humans. The officially accepted value for the energy content of gum acacia has varied between zero and 100% of the bomb calorimeter value. The currently accepted value is 50% of the bomb calorimeter value. Some food materials pass out of the body unmetabolised.

In recent years obesity has become a major problem. Consequently, the aim of some food manufacturers has shifted to producing products with a lower energy content. The energy content of a food can be reduced by using non-metabolised ingredients, reducing the density, or replacing high energy density ingredients with lower energy density ingredients. An example would be replacing fat with anything else.

2.15.1.1 Special Nutritional Needs. Several groups have special nutritional needs, *e.g.* children, the elderly, pregnant women, manual workers and athletes. There are groups who have medical special needs such as those with allergies or diabetes. The latter groups need special foods that are, sometimes, outside ordinary food regulations and, therefore, have special exemption from them.

If more energy is expended than is consumed then the energy will be provided from the body's food stocks. This will obviously cause the subject to lose weight. A vast industry has grown from this simple observation – the slimming business. Slimming diets abound, in some cases supported by books, clubs and special foods.

2.15.2 Food Groups

The content of foods is normally presented in terms of the following components.

2.15.2.1 Proteins. These compounds are essential to build and maintain muscles. Any protein above this requirement is metabolised.

2.15.2.2 Carbohydrates. These compounds contain only carbon, hydrogen and oxygen. The commonest carbohydrate is starch. Carbohydrates are a source of energy.

Some compounds that only contain carbon, hydrogen and oxygen are not classified as carbohydrates. One example is ethyl alcohol.

2.15.2.3 Fats. Oils are merely fats that are liquid at room temperature. Nutritionally they are all considered to be fats. Chemically, fats are esters that can be formed from fatty acids and the trihydric alcohol glycerol. Lipid chemists now call them triacylglycerols, while nutritionally they are still known by the older name of triglycerides. The properties of a fat depend on the fatty acids present. The important factors, nutritionally, are the degree of unsaturation of the fatty acids, the chain length and, for unsaturated fatty acids, whether the double bonds are cis or trans. The major function of fat is as a source of energy. However, a few fatty acids have a special nutritional function and are known as essential fatty acids since the human body needs them but cannot synthesise them. These essential fatty acids are divided in to the two classes of omega 3 and omega 6. The omega 6 acids are found in vegetable oil.

Nutritional understanding of the effect of fats in the diet has made considerable progress.[26–30] It was understood that saturated fats (see Chapter 3, Section 3.8) were the least beneficial as they raised serum cholesterol. High serum cholesterol is now associated with heart attacks and strokes. There was for this reason pressure over the cholesterol content of foods. This pressure has now been relieved since it appears that dietary cholesterol is not a particularly serious issue. The human body makes cholesterol, so dietary cholesterol does not necessarily affect serum cholesterol level as dietary intake can be compensated for by reduced cholesterol synthesis.

Unfortunate individuals with the highest serum cholesterol have an inherited metabolic defect such that their bodies cannot metabolise cholesterol. These individuals need drug treatment not dietary measures.

Saturated fats were found to elevate serum cholesterol levels while polyunsaturated fats reduce them. This led to advice to consume less saturated fats and more polyunsaturated fats.

Polyunsaturated fats are relatively rare and the main available source is sunflower oil. They also have the disadvantage that their high degree of unsaturation also makes them unstable on cooking and storage. If sunflower oil is used for frying the residues will eventually polymerise. Animal testing has revealed that such residues are potentially carcinogenic, leading to advice that sunflower oil used for frying should be discarded after use.

Further research on serum cholesterol revealed that is exists in two forms, high and low density lipoprotein. The high density lipoprotein appears to consist of cholesterol that is being moved to the liver for metabolism while low density lipoprotein appears to be cholesterol that is likely to block arteries. This has led to the labelling of high density lipoprotein as good cholesterol and of low density lipoprotein as bad cholesterol.

Saturated fat raises the level of both sorts while polyunsaturated fats lower both types. Further research revealed that monounsaturated fats selectively elevate the high density cholesterol. Still further research has revealed that trans fatty acids elevate the low density lipoprotein and reduce the high density lipoprotein levels. Trans fatty acids are rare in nature, occurring mainly in milk; they are, however, produced in the hydrogenation of oils. The association of trans fatty acids with elevated risks of heart disease is such that recommendations have been issued that not more than 5% of dietary energy should come from trans fatty acids. Some countries have already legislated to restrict trans fatty acid levels in foods. These findings have forced the oil and fats business to abandon the use of hydrogenation.

Fats in general have been criticised for being high in energy – at 9 kcal g^{-1} they have over twice the energy density of carbohydrates or proteins (4 kcal g^{-1}). Apparently, the easiest way of reducing the energy content of a food is to replace the fat with another ingredient. Technically, fat makes the texture of foods more pleasant. Some attempts at producing very low fat content foods led to a product that can be described as like chewing a doormat. Some low fat foods rely on using a small quantity of emulsifier to replace a large quantity of fat. Curiously, as fats tend to be more expensive than other ingredients commercial pressures operate in the same direction.

2.15.3 The Glycemic Index

An important concept in understanding how foods affect the human body is the glycemic index. For many years it had been felt that a diet that was high in refined foods was unhealthy. The results obtained in various studies were inconsistent. The technology that made progress possible was the measurement of blood glucose. The glycemic index compares the effect on the blood glucose of a given food against that produced by the same quantity of dextrose. Confusingly, an alternative version uses white bread as the control sample.

The glycemic index has produced some unexpected results. Mashed potato produces a much bigger surge in blood glucose than an equivalent quantity of sucrose. Chemically, this is understandable, as potato starch is a polymer of dextrose. The body can breakdown potato starch, quickly liberating the dextrose which is rapidly absorbed. Sucrose is a disaccharide composed of fructose and dextrose. When the sucrose has been split, also referred to as inversion, into the monosaccharides the fructose is metabolised independently of insulin. Only the dextrose derived from sucrose can affect the blood sugar levels. The glycemic index of a food depends on the nature of the carbohydrates present, the fat content and whether dairy ingredients are present. Both fat and dairy ingredients slow down absorption.

In the case of complex carbohydrates the glycemic index is a measure of how quickly the carbohydrate is broken down. An example is to compare potato starch and polydextrose. Both are polymers of dextrose but potato starch in the form of mashed potato is rapidly broken down and causes a surge in blood sugar, *i.e.* it has a high glycemic index. In contrast, polydextrose, which has the dextrose units linked $1 \rightarrow 6$, a link that is rare in nature, is only 25% metabolised and has a very low glycemic index.

2.15.4 Trace Elements

Several trace elements are essential for a healthy life. One example is iodine, which is needed to make the thyroid hormone. An iodine deficiency leads to goitre. This disease was once called Derbyshire throat as it was once common in that county. Goitre was common in inland areas where the soil is low in iodine and access to seafood is poor.

While some trace elements are essential for a good health an excess is undesirable. This is particularly so with elements that are low in solubility as the body is unable to remove the excess.

Selenium is an essential trace element for optimal health but an excess is toxic. The British population is now reckoned to be selenium deficient because more European wheat is now used to make bread at the expense of North American wheat. The difference in the selenium content is caused by the difference in the selenium content of the soils. In geological terms, European soil is older and the selenium has washed out.

Modern analytical methods are so sensitive that low levels of many elements that would be toxic at higher levels can be found, *e.g.* arsenic. At one time some US states had laws banning any food with measurable quantities of arsenic. At this time, lead arsenate was used as an insecticide – if arsenic was present in measurable quantities then insecticide residues were present. However, such low levels of arsenic are not a health problem.

2.15.5 Vitamins

The term vitamin is a misnomer, the name means vital amines, and while vitamins are essential for life they are not, as was originally supposed, amines. Most vitamins were discovered as a result of a deficiency disease produced by a restricted diet. Long voyages on sailing ships with a diet composed of ship's biscuit, dried beans, dried peas and salted meat produced scurvy. In the worst cases the whole crew were affected, but the ship's officers tended to be less severely affected.

Ships that carried beer tended to be less affected than those that carried water and spirits. Presumably, the beer contained some vitamin C, possibly from the habit of "dry hopping", *i.e.* adding a few hop cones to each barrel. Eventually, it was found that lemon or lime juice every day could prevent scurvy. The admiralty waited fifty years before they applied the discovery and then insisted that all British ships carried lime juice.

While it was known that lime or lemon juice would prevent scurvy the active ingredient remained elusive. Experiments were conducted using dilute hydrochloric acid as a substitute. Predictably, they did not work.

The structure of ascorbic acid was not known until the 1930s. The only knowledge of the appropriate dose came from the amount required to prevent scurvy. It remained unknown whether a higher dose was beneficial.

This uncertainty at one point led to the American government having daily requirements for vitamins that were twice those advised by the British government.

Claims were made of the benefits of very high doses of vitamin C. It has transpired that no discernible benefits seem to occur. The only effect of extremely high doses of vitamin C is a laxative one.

While the human body can remove an excess of any water-soluble vitamin, excesses of fat-soluble vitamins are more serious. Early arctic explorers discovered that the Inuit regarded seal liver and polar bear liver as taboo and must not be eaten. Those explorers who ignored this advice risked retinol poisoning as the livers of both these species are rich in retinol (vitamin A) that can not be excreted. The effects of retinol poisoning are extremely unpleasant. It is for this reason that fortification with fat-soluble vitamins is not likely to be undertaken.

2.15.6 Nutritional Labelling

This is an area that is in a state of flux at the time of writing. There is increased pressure from government sources aimed at trying to reduce obesity and heart disease. The EU has also become involved. Indeed, the recent additions to the EU make it particularly difficult to label products with small packages in all EU languages.

2.15.6.1 The Current Position. This could soon change and readers are advised to check these points on the food standards agency website (www.foodstandards.gov.uk).

Unless a specific nutritional claim is made the provision of nutritional information is voluntary.

Around 80% of pre-packaged foods carry some nutritional information and failure to do so is likely to lose sales. There will be those customers who have medical advice about diet coupled to those who are trying to lose weight. Both classes of customers are likely to ignore products that do not have a nutritional label. Indeed, some may view the omission as an attempt to cover up an unfavourable situation.

2.15.6.2 The EU Influence. The EU has recognised that different nutritional labelling requirements of member states could constitute a barrier to cross border trade. Directive 90/496/EEC was adopted to this end in 1990 to harmonise nutritional labelling. While this directive does not make nutritional labelling compulsory, except where a claim is made, it does prescribe the format to be used.

2.15.6.3 Formats. Information should be provided in one of the following formats:

Group 1 information: energy, protein, carbohydrate and fat.
This format is referred to as the big 4
Group 2 information: energy, protein, carbohydrate, sugars, fat, saturates, fibre and sodium (this order is specified).
This format is known as the big four plus the little four.

In either case the quantities must be given either (1) per 100 mL or per 100 g or (2) per 100 mL or per 100 g and per serving. NB There is not an option of just giving the information per serving. All this information should be supplied in one place in a tabular format with the numbers aligned if space permits.

Declarations may only be made regarding vitamins and minerals that are listed in the annex of the directive provided that the vitamins and minerals are present in "significant amounts". "Significant amounts" are defined as that per 100 g or 100 mL of the food, or per package – if the package contains one portion it should contain 15% or more of recommended daily amount as listed in Table 2.

Declarations may be made regarding the following:

- starch
- polyols
- mono-unsaturates
- polyunsaturates
- cholesterol

However, if declarations are made about polyunsaturates, monounsaturates or cholesterol the amount of saturates must also be given. Here, science has moved on from the time when saturated fats were the fats to avoid to advice to avoid trans fatty acids and to prefer monounsaturated fat to polyunsaturated fats.

The regulations prescribe the energy values to be used as in Table 3. There are agreed values for some other materials. The synthetic dextrose polymer polydextrose is accepted to be only 1 kcal g^{-1} (or 4 kJ g^{-1}) even though it would otherwise fall within the definition of a carbohydrate. The accepted value for gum acacia, a polysaccharide obtained from trees of the species *Acacia senegal* and closely related species, is 2 kcal g^{-1} (or 8 kJ g^{-1}).

Table 2 *Recommended daily amounts of vitamins and minerals*

Vitamin or mineral	*Recommended daily amount*
Vitamin A	800 μg (micrograms)
Vitamin D	5 μg
Vitamin E	10 μg
Vitamin C	60 μg
Thiamin	1.4 mg (milligrams)
Riboflavin	1.6 mg
Niacin	18 mg
Vitamin B6	2 mg
Folacin	200 μg
Vitamin B12	1 μg
Biotin	0.15 mg
Pantothenic acid	6 mg
Calcium	800 mg
Phosphorus	800 mg
Iron	14 mg
Magnesium	300 mg
Zinc	15 mg
Iodine	150 μg

Table 3 *Prescribed energy values*

Substance	*Calorie value (kcal g^{-1})*	*Energy in joules (kJ g^{-1})*
Carbohydrates except polyols	4	17
Polyols	2.4	10
Protein	4	17
Fat	9	37
Ethanol	7	29
Organic acid	3	13

While the value of 2.4 kcal g^{-1} (10 kJ g^{-1}) is accepted throughout the EU the same value is not accepted everywhere for all polyols. At one point, the accepted value for one polyol differed by a factor of two between Canada and the USA.

The value to be used for the protein content is defined as the Kjeldahl nitrogen multiplied by 6.25. Similarly, there are definitions regarding saturated fatty acids.

2.15.6.4 Average Values. A major issue is how to cope with the normal variation in the composition of raw materials. The directive defines an average value as "the value which best represents the amount of the nutrient which a given food contains and reflects allowances for seasonal variability, patterns of consumption and other factors which may cause the actual value to vary".

2.15.6.5 The Declared Values. These values should be the average values as defined above and should be derived from:

- The manufacturer's analysis of the food.
- Calculation from the known or actual average values of the ingredients used.
- Calculation from generally established and accepted data.

Data from McCance and Widdowson's *The Composition of Foods* could satisfy the second and third case.[31]

While the declaration should be for the food as sold, data can be given for the food after preparation if instructions for preparation are given. A nutritional analysis of unshelled walnuts has a limited value for example.

2.15.6.6 Nutritional Claims. The directive defines a nutritional claim as follows:

> Any representation and any advertising message which states, suggests or implies that a food stuff has particular nutritional properties due to the energy (calorific value) it provides, provides at a reduced or increased rate or does not provide and/or due to the nutrients it contains, contains in reduced or increased proportions or does not contain.

2.15.6.7 Differences Between EU and British Law. One problem with EU and British law is an important difference in approach. British law has always worked on the basis of that which is not illegal is automatically legal while EU law takes the approach that what is not expressly permitted is illegal. This has ended up with many British regulations running "No person shall . . . ".

Dietary Fibre. An area of some contention is what should be included as dietary fibre. There are several different methods of measuring dietary fibre and all of them will give different results on the same sample. The EU favours the Association of Analytical Chemists (AOAC) method, which includes lignin and resistant starch, while the UK has preferred the Englyst method and has defined fibre as non-starch polysaccharides from cell walls as the only substances that count as fibre.

When dietary fibre became a nutrition issue it was assumed that the insoluble fibre such as bran was the most beneficial subcomponents. It now appears that soluble fibres can lower blood cholesterol while insoluble fibre such as wheat bran merely speeds up the transit of food through the gut. Developments in this area are awaited.

2.15.6.8 Other Countries. While the EU may have standardised nutritional statements throughout the EU the rest of the world has different rules. The UK and the USA might share a common language but they do not share a common nutrition statement.

2.15.6.9 The Consumer. An important part of the public information process is the consumer. All the nutritional information in the world is no use unless the consumers can make use of it. While the consumer does not need to know the chemical definition of trans fatty acids it is to be hoped that they can understand the need to limit the consumption of such acids.

Energy values can be a problem as one survey of opinion found that consumers thought that calories were bad but energy was good, thus negating an important thermodynamic principle. It has taken so long to educate the consumer that the calorie value is the important item in describing the potential of a food to lead to obesity that the prospect of changing to joules is not good. Unfortunately, the joule value is always higher than the calorie value. The other unfortunate thing is the use of the large calorie, *i.e.* a kilocalorie, written as Calorie. This usage will take a long time to fade away.

Labels have to avoid being misleading to the consumer. As an example some academic chemists dislike the practice of referring to substances such as sorbitol as polyols and would prefer them to be called sugar alcohols. The confusion that this would cause would be very unfortunate. Some of those who need to control their intake of small carbohydrates would be wondering if these substances are sugar or not?

2.16 FOOD ALLERGY AND INTOLERANCE

Of necessity, anyone working in the food industry has to be aware of food allergies. They involve one of two different immunologic mechanisms. The most common one involves globulins E (IgE). The less common type is the cell-mediated food allergy.

2.16.1 Immunoglobulin E (IgE)-mediated Food Allergies

These are allergies where immunoglobulin E antibodies can be detected. These allergies tend to be rapid in onset with a time from consumption to symptoms ranging from minutes to a few hours.

IgE antibodies are very highly specific. In a few cases they will cross react with closely related substances but generally they will not.

The first stage of an IgE allergy is a sensitisation process where mast cells and basophils, which are specialised cells in the blood, become

Table 4 *Symptoms of IgE allergy*

Gastrointestinal symptoms	*Respiratory symptoms*	*Cutaneous symptoms*	*Other symptoms*
Vomiting	Rhinitis	Urticaria (hives)	Anaphylactic shock
Diarrheoea	Asthma	Dermatitis	Laryngeal edema
Nausea		Angiodema	

sensitised. Initially, the IgE antibodies are produced by plasma cells and attach themselves to the outer membrane surface of the mast cells and the basophils. After this these cells have been sensitised to the allergen. If there is no further exposure there will be no problem.

If there is a further exposure then the allergen crosslinks two IgE molecules, which triggers the release of histamine, prostaglandins and leukotrienes. These then produce the allergic response.

The response varies, depending on the individual and the allergen. The worst case is that anaphylactic shock arises when gastrointestinal, cutaneous and respiratory symptoms occur in conjunction with a dramatic fall in blood pressure and cardiovascular complications. Death can occur within minutes of anaphylactic shock. Table 4 lists the other symptoms of IgE allergy.

IgE allergies can be detected in two different ways. The more modern is the radioallergosorbent test (RAST). Here, a small sample of the patient's blood is collected and the serum is tested for the binding of IgE to food protein bound to a solid matrix using radiolabelled or enzyme-linked antihuman IgE.

The older test is the pin prick test where a small quantity of the antigen is placed under the skin. If there is an allergic reaction a wheal and flare, *i.e.* a hive, will appear. The pin prick test is not suitable for those who undergo anaphylactic shock.

2.16.2 Cell-mediated Food Allergies

These are called delayed hypersensitivity reactions since they normally occur 6–24 hours after exposure. A cell-mediated allergy involves the interaction of food allergens with sensitised lymphocytes, which usually occurs in the gastrointestinal tract. The sensitised lymphocytes produce lymphokines and the generation of cytotoxic T lymphocytes. These latter cells destroy other intestinal cells, including the epithelial cells that are critical for absorption.

Coeliac disease, which is an allergy to gluten (specifically the gliadin fraction), is a cell-mediated food allergy.

Symptoms of coeliac disease are diarrhoea, bloating, weight loss, anaemia, bone pain, chronic fatigue, weakness, muscle cramps as well as failure to gain weight and growth retardation in children.

2.16.3 Food Intolerance

Food intolerance occurs when a food ingredient cannot be metabolised. The commonest food intolerance is lactose intolerance. This occurs when the enzyme lactase is absent. This happens when humans do not consume milk after weaning; a state which is the norm in most of Asia. Thus it is possible that the bulk of the world's population is lactose intolerant.

Some individuals have convinced themselves that they have a sensitivity to some food or other which might or might not be true.

2.17 THE SCIENCE OF AERATED PRODUCTS

All aerated products are technically foams. One point of view is that aerated products are a way of making the public pay for air. This is not true in the case of many bakery products such as bread as the product is not sold by volume but by weight. The only food where this claim could be true is ice cream which is sold by volume not weight. Not only does the consumer pay for the air in ice cream as ice cream is subject to standard rate value added tax the consumer pays VAT on the air as well!

The two problems in making an aerated product can be simply stated as making the bubbles and stabilising them when they have been made.

2.17.1 Making the Bubbles or Leavening

The following methods are all used.

- Mechanically beating the product to produce a foam. An example of this is whisking an egg into a foam with flour, as in the production of sponge cakes.
- Steam produced by heating water-containing materials can also be used, *e.g.* Scandinavian flat breads and puff pastry.
- Yeast is the leavening agent in bread and yeast raised wafers. Carbon dioxide produced by the yeast not only causes bubbles to form but also dissolves, reducing the pH. The other major product of yeast action, ethyl alcohol, boils away during the cooking phase.
- Chemical leavening takes place when carbon dioxide is produced chemically rather than biochemically. Typically, this involves treating sodium bicarbonate with an acid such as sodium acid

phosphate, mono calcium phosphate, sodium aluminium phosphate or glucono-δ-lactone. One hundred grams of baking powder generates 8.2 litres of carbon dioxide, which is 340 nM and weighs 15 mg. Chemical leavening is the usual method in biscuits and cakes. It can be used for wafers and is sometimes used for pizzas although authentic pizzas are yeast raised.

2.17.2 Stabilising the Foam

Stabilising the foam is the more difficult part of the process. Some systems such as egg albumen will foam easily and set on cooking, which are desirable characteristics in the food industry.

The important issue is to produce a layer between the bubbles that is sufficiently stable that the bubbles do not burst either before the food is cooked or served. In bakery products the foam has to be stable until the surrounding proteins have been denatured and the starch gelatinised.

Some factors such as the viscosity of the medium can make the foam harder to aerate but when aerated they work to resist the collapse. Other factors normally only work in one direction.

A stable foam is likely to have ingredients that are in a low energy state at the air–liquid interface. Substances that fit this description include proteins, emulsifiers some fats and fat components such as diglycerides monoglycerides and fatty acids. Food law uses the term emulsifier and stabiliser to cover the situation where the ingredient is stabilising an emulsion rather than helping to form it.

The layers between bubbles can be as thin as one molecule thick. Some systems, usually those involving eggs, are fairly robust. Others can easily be caused to collapse. Some years ago purified lactalbumin was offered as a substitute for egg albumin. The purified lactalbumin would foam up when whisked but the foam would collapse if more than a trace of fat was present. While there might be food systems where this would be acceptable it is not generally acceptable as a general replacement for egg albumen.

Presumably, when fat was present it was drawn into the layer between the bubbles and disrupted the protein layers, causing the foam to collapse.

It is possible to write about designing a system to produce a stable foam on the basis of accumulated scientific knowledge. However, notably, the vast majority of food systems were the product of empiricism and serendipity! It remains to be seen if advances in scientific understanding can produce any food foam that is as successful as a meringue.

2.17.3 Fat in Bread

While the destabilising influence of fat on some food foams has already been mentioned there are food products that can be made both with and without added fat. One of the most important examples is bread. Adding fat increases the shelf life, presumably by inhibiting starch retrogradation. It is possible to compare the behaviour of the same bread with and without added fat. In Junge and Hoseney, samples of the same dough but with and without added fat were placed in a resistance heating oven.[32] Both doughs expanded at the same rate until they reached 55°C. The fat-free dough continued to expand but at a reduced rate. The fatted dough continued to expand at the same rate until it reached 75°C. Thus the addition of 3% of fat produced a loaf around 100 mL bigger. Similar results were obtained by using 0.6% monoglyceride or 0.5% sodium stearoyl lactylate.[33] The effect of these small quantities of emulsifiers is impressive. The mechanism is unknown.

REFERENCES

1. L.C. Maillard and M.A. Gautier, Action des acides amines sur les sucres: formation des melanoidines par voie methodique, *Compt. Rend. Acad. Sci.*, 1912, **154**, 66–68.
2. C. Fisher and T.R. Scott, *Food Flavours Biology and Chemistry*, The Royal Society of Chemistry, Cambridge, 1997.
3. J. O'Brien, H.E. Nursten, M.J.C. Crabbe and J.M. Ames, (eds.) 1998. *The Maillard Reaction in Foods and Medicine.* Royal Society of Chemistry, Cambridge.
4. S. Cauvain and L. Young, in *Baking Problems Solved*, Woodhead Publishing Ltd. Cambridge England 2001 p136.
5. C.H. Chen and W. Bushuk, *Can. J. Plant Sci.* 1970, **50**, 9–14.
6. M.C. Gianibelli, O.R. Larroque, F. MacRitchie and C.W. Wrigley, aaccnet.org/cerealchemistry/freearticle/gianibelli.pdf.
7. C.W. Wrigley, H.P. Manusu, S. Paranerupasingham and F. Bekes, in *Gluten 96*, ed. C.W. Wrigley, published by Royal Australian Chemical Institute Melbourne, Australia.
8. F. MacRitchie, *Cereal Foods World*, 1980, **25**, 382.
9. M. Byers, B.J. Miflin and S.J. Smith, *J. Sci. Food Agric.*, 1983, **37**, 447–462.
10. G. Danno, *Cereal Chem.*, 1981, **58**, 311–313.
11. A. Graveland, P. Bosveld, W.J. Lichtendonk, J.H.E. Moonen and A. Scheepstra, *J. Sci. Food Agric.*, 1982, 1117–1128.

12. A. Graveland, P. Bosveld, W.J. Lichtendonk, J.P. Marseille, J.H.E. Moonen and A. Scheepstra, *J. Cereal Sci.*, 1985, **3**, 1–16.
13. N.K. Singh, G.R. Donovan, I.L. Batey and F. MacRitchie, *Cereal Chem.*, 1990, **67**, 150–161.
14. N.K. Singh, G.R. Donovan, I.L. Batey and F. MacRitchie, *Cereal Chem.*, 1990, **67**, 161–170.
15. T. Burnouf and J.A. Bietz, *Cereal Chem.*, 1989, **66**, 121–127.
16. N.K. Singh, K.W. Shepherd and G.B. Cornish, *J. Cereal Sci.*, 1991, **14**, 203–208.
17. R.B. Gupta and F. MacRitchie, *J. Cereal Sci.*, 1991, **14**, 105–109.
18. J.A. Bietz, *Baker's Dig*. 1984, **58**, 15–17, 20–21, 32.
19. I.L. Batey, R.B. Gupta and F. MacRitchie, *Cereal Chem.*, 1991, **68**, 207–209.
20. R.B. Gupta and K. Shepherd, *Theor. Appl. Genet.*, 1993, **80**, 65–74.
21. J.-L. E. Lew, D.D. Kuzmicky and D.D. Kasarda, *Cereal Chem.*, 1992, **69**, 508–515.
22. F.R. Huebner and J.A. Bietz, *Cereal Chem.*, 1993, **70**, 506–511.
23. B.X. Fu and H.D. Sapirstein, *Cereal Chem.*, 1996, **73**, 143–152.
24. R. Tkachuk and K.H. Tipples, *Cereal Chem.*, 1966, **43**, 62–79.
25. Peat *et al.*, *J. Chem. Soc.* 1952, 4546.
26. A. Ascherio, M.B. Katan, P.L. Zock, M.J. Stampfer and W.C. Willett, Trans fatty acids and coronary heart disease (sounding board), *N. Engl. J. Med.*, 1999, **340**(25), 1994–1998.
27. T. Byers, Hardened fats, hardened arteries? *N. Engl. J. Med.*, 1997, **337**, 1544–1545.
28. F.B. Hu, M.J. Stampfer, J.E. Manson, E. Rimm, G.A. Colditz, B.A. Rosner, C.H. Hennekens and W.C. Willet, Dietary fat and the risk of coronary heart disease in women, *N. Engl. J. Med.*, 1997, **337**, 1491–1499.
29. A.H. Lichtenstein, Trans fatty acids and hydrogenated fat, *Nutrition Today*, 1995, **30**(3), 102–107.
30. L. Liten and F. Sacks, Trans-fatty acid content of common foods, *N. Engl. J. Med.*, 1993, **329**, 1969–1970.
31. *McCance and Widdowson's The Composition of Foods*, 6th Summary Edition, The Royal Society of Chemistry, Cambridge, 2002.
32. R.C. Junge and R.C. Hoseney, A mechanism by which shortening and certain surfactants improve loaf volume in bread, *Cereal Chem.*, 1981, **58**, 408–412.
33. W.R. Moore and R.C. Hoseney, Influence of shortening and surfactants on retention of carbon dioxide in bread dough, *Cereal Chem.*, 1986, **63**, 67–70.

CHAPTER 3

Raw Materials

3.1 GRAINS

Whole grains seldom form an ingredient of products directly. When they do it is usually as a decoration. However, the properties of important ingredients such as flour depend critically on the properties of the grains from which they are made.

3.1.1 Wheat

Wheat is the major source grain for bakery ingredients. The cultivation of wheat is the basis of western civilisation. Botanically wheat is a member of the grass family (Grammacidae). Bread making depends on the proteins in wheat.

Wheat is normally categorised into hard and soft wheat as well as bread making and non-bread making. Almost all hard wheats are of bread making quality while some soft wheats do have bread making quality while others do not. French bread, for example, is based entirely on soft wheat. Commercially, bread-making wheats are of higher value than non-bread making ones. Acceptance for bread making requires the wheat to pass tests on protein quantity, protein quality and falling number. In the UK and some other countries soft wheat is grown either for use in products like biscuits or for animal feed. Biscuits are not made from soft wheat flour because it is cheaper but because high protein hard wheat flour will not work. Specifications for wheat for uses such as biscuits tend to be much looser than for bread making. In practice, the baking industry traditionally paid little attention to the variety of wheat used for biscuit flours. Soft wheat varieties are selected for their growing properties rather than for the sort of testing that is undertaken for bread-making wheat. Increasing automation in biscuit and wafer making factories has made them less tolerant of the flour that is used, so improvements in production have led to a stricter specification for raw materials.

An important factor in the bread making quality is the weather. The uncertain British climate often produces widely varying wheat from year to year. The bread making quality of wheat is enhanced by hot dry weather immediately before harvest. In 1987 the weather in East Anglia, which is normally the major wheat producing region of the UK, was such that the winter wheat was all of too low a falling number to be any use for making bread. The spring sown wheat in the same region was of excellent quality. The hot summer of 1976 gave excellent bread making wheat but caused problems for biscuit manufacturers because the soft wheat varieties had too much protein.

The National Association of British and Irish Millers (NABIM) classifies British wheat into four groups. Group 1 wheats all have good bread making potential; they are all hard wheats. Group 2 wheats have some bread making potential but are not necessarily as consistent as group 1. Group 3 wheats are soft wheats with the extensible sort of dough properties that make them suitable for biscuits and cakes. Group 4 wheats have little use in foods and are normally used as animal food.

3.1.2 Barley

Barley is not suitable as an ingredient for bread; it does, however, form a bakery ingredient as malt. Malting is the process where the grain is steeped in water to start it sprouting; the grain is then kilned to arrest the process. Malting causes the conversion of some of the starch into sugars. The major food use of malt is in the manufacture of beer and whisky.

Malt is used in bakery products in the form of whole grains, malt flour and malt extract. Malt extract is a sticky brown syrup made by saccharifying malt. The syrup has a sufficiently high concentration to be stable against microbial spoilage. A wide range of malt extracts are made, depending in their intended use. Bakery syrups are normally graded on their colour and are processed so that they have no residual diastatic activity as this would break starch down into sugars, thereby lowering the falling number. Brewery syrups are normally diastatic so that they can convert unmalted malted grains such as wheat flour, maize flakes or rice flakes into sugars.

While malt flour is not suitable for bread making it is added to wheat flour in small quantities to feed the yeast, open out the texture and improve the flavour. This practice has declined in recent years for several reasons. When British bread flour was mainly made from Canadian wheat with a Hagberg Falling Number of around 600 the addition of some malt flour would open the texture of a loaf by attacking some of the starch. The resulting sugars would feed the yeast while some of the reducing sugars produced would react with proteins in the Maillard reaction, improving the flavour and the colour of the crust.

An overdose of malt flour will introduce too much enzyme activity in the dough, potentially reducing the product to a sticky syrup. Fungal α-amylase is now much more commonly used than malt flour. However, some malt flour is still used, particularly in wholemeal bread.

3.1.3 Rye

Rye, like wheat, has bread making potential and rye dough can develop; however, rye bread does not have the potential for expansion that wheat bread does.

Rye will grow under much harsher conditions than wheat. In Germany and Scandinavia rye breads, either wholly rye or mixed with other grains, are eaten. In the UK, rye is little used and the somewhat bitter flavour that it imparts is in general not appreciated. The small amount of rye used is deployed in health food products, crisp breads and in making rye products to satisfy foreign tastes.

3.1.4 Maize

This grain, botanically *Zea mays*, is the native grain plant of the Americas where it is known as corn. Sometimes it is referred to as *Indian corn*.

Maize has some potential as a bread making ingredient and is so used in the places where it is grown. Presently, maize is not a viable grain crop in Europe owing to the climate.

Imported maize is the raw material for several food ingredients used in the bakery industry. While maize can be dry milled like wheat, it is more commonly wet milled. The wet milling process is much better suited to separating the different components of maize so that the oil, the protein and the starch can be recovered separately. Maize starch is used directly in bakery products as corn flour, so-called even in the UK.

3.1.5 Dried Gluten

Dried gluten is the protein fraction from wheat that has been milled to flour and the flour kneaded with water. This process allows the insoluble starch to be removed and the resulting proteinaceous mass is dried to produce dried gluten. Obviously, the flour should be milled with as low a starch damage as possible.

3.1.6 Soy Beans

Soy beans are another crop that will not grow in Europe. The soy bean is used as a source of both protein and vegetable oil. Enzyme active soy flour has been used in bread since the 1930s. The flour contains a lipoxygenase system that assists with the development of the dough and slightly bleaches the bread. Soy flour is classed as an ingredient rather than an additive.

The enzymes present in soy bean flour are not desirable in all products because they interact with the fat phase and can cause "beany" off-flavours. Enzyme inactive soy flour is made for these products.

Soy bean oil is suitable for use in frying or as a salad oil. It is also a popular starting point for making margarine.

3.1.7 Margarine

Margarine was always intended as a substitute for butter. Although originally made from beef tallow, the main raw material for many years has been vegetable oil. To make a product with a similar consistency to butter most vegetable oils have to be hardened. This was normally done by hydrogenating the vegetable oils over a catalyst. Unfortunately, partial hydrogenation produces trans fatty acids, which are unhealthy and have already been restricted in some countries.

At present, margarine producers are moving to use fractionation and interesterification to produce the required properties. A new technology uses lipase enzymes to rearrange fatty acids in a controlled way.

Margarine has always had the advantage over butter in that the properties of the product can be tailored to give the best performance in a particular system. For puff pastry, specialised margarines are easier to work with than butter.

3.2 MILLING

Milling converts grain into flour. Grinding grain between two stones is a process of great antiquity. Pairs of hand operated stones known as quorns have been found among the relics of some of the earliest agrarian settlements. Using water power to drive a pair of stones must have made for a much easier life. Driving the stones by wind power as a windmill is a later development introduced into Europe as a consequence of the crusades. A later innovation was the roller mill. In this system the wheat berry is dismantled layer by layer. In a stone mill the wheat berry is simply crushed, which produces a wholemeal flour – any other type of flour can only be made by performing a separation on the flour.

In a roller mill wholemeal flour can only be made by recombining all the fractions at the bottom of the mill. This is why some wholemeal flour is still stone ground. The stone mills used are not antiques but are usually driven by electricity. There is no reason why roller mills or stone mills should not be driven by water power in the form of a water turbine.

The important point for the baker is that milling affects the flour in ways that cannot be altered elsewhere in the process. The most important effect is the amount of starch damage produced in the milling process. Damaged starch can absorb water while undamaged starch can not. Because damaged starch can be attacked by amylase enzymes it is a source of food for the yeast. The action of amylase enzymes on damaged starch softens the dough. While this may be desirable in some products, if it goes too far the dough will become sticky and unhandleable. In bread flours based on hard wheat a high degree of starch damage is desirable since it allows a greater yield of bread as the starch absorbs water. Excessive amounts of starch damage lower bread quality. In some soft wheat flours, as used in biscuits, low starch damage is desirable.

After the milling process, any gaseous treatments are applied, any powder treatments, *e.g.* ascorbic acid, are added, as well as any fortifying ingredients such as calcium sulfate. Different countries have various policies on fortifying flour. In the UK, white flour is fortified with calcium to make up for the calcium lost by not making a wholemeal flour. In the USA, bread is fortified with folic acid. It is possible for an untreated flour to be mixed with a flour improver containing the powder treatments.

One property of the flour that is controlled by the miller is the extraction rate. Wholemeal flour has a 100% extraction rate, with brown, white and patent white having progressively reduced extraction rates. One obvious difference is the colour. Another is that the quality of the protein increase towards the middle of the wheat berry from which patent flour is produced. Thus, patent flour is sometimes used not to produce whiter bread but in products like filo pastry or West Indian patties where the strength that patent flour gives is important and the colour is irrelevant.

Wheat is normally dry milled. Maize can be dry milled but is normally wet milled. The wet milling process allows the maize to be fractionated into starch protein and oil.

Other additions to flour are fungal α-amylase malt flour, malt grains or other mixtures of grains. Alternatively, the flour can be supplied untreated by the miller and any treatments needed can be applied in the bakery. This could be an attractive option for a bakery that made a mix of products, some of which needed untreated flour while others, *e.g.* bread, needed flour treatment.

3.3 GRADES OF FLOUR

Flour of course is the main ingredient in baked goods. Flour is normally supplied to meet a specification. In some cases the specification is very wide or it is very tight. Some specifications ensure that the flour is fit for making a specific product, *e.g.* bread.

Most millers supply a range of flours to bakers. Some millers specialise in producing niche products. A typical range of bread flours would look as follows.

3.3.1 Top Grade

This would have a substantial proportion of Canadian wheat and would be suitable for any long process. It can be used to make ordinary bread but is more likely to be used to make Viennas or rolls. The protein content could be as high as 14% with a water absorption of 62–64%. This product would have a high tolerance in the bakery. One use would be with a suitable improver to produce very well blown up rolls.

3.3.2 Baker's Extra Grade

This grade has less imported wheat and a lower protein content than the top grade but more than the baker's grade. The protein content might be 13%.

3.3.3 Baker's Grade

This is the standard grade used by small bakers to make bread. There will be sufficient third country wheat, probably Canadian, for it to work in a long process such as bulk fermentation. The protein content would be around 12%.

3.3.4 Euro Baker's Grade

This product is really a creature of the EU's Common Agricultural Policy (CAP). It would be a baker's flour similar to the standard baker's grade but without any non-EU wheat. As it would match the protein content of baker's grade, dried vital wheat gluten would be used to make up the protein content. This product would only be suitable for quick processes like the spiral mixer. Not all millers would produce this grade. Changes to the CAP may destroy the commercial viability of this product.

3.4 TYPES OF FLOUR

3.4.1 Chorleywood Bread Flour

This is the flour that would go into the Chorleywood bread plants. It would be based on all EU wheat (in most years all English). The protein content would probably be 10.6–11.5%. This flour could also be used for making puff pastry.

3.4.2 Patent Flours

All of the above flours are white flours of ordinary whiteness. If the extraction rate is reduced still further, whiter flour known as patent flour is obtained. A patent flour can be produced from the grist of baker's grade or higher flours. The resulting flour will not only be whiter it will have a lower protein content. The quality of the protein will be higher.

Patent flour has two classes of use. It can be used to make whiter bread or where very high protein content is required. The use of patent flour to make bread seems to be dying out. Its use does, however, remain popular in South Wales. There are various examples of products where patent flour is used for its protein quality, *e.g.* filo pastry and West Indian patties. Both of these products are brown so the colour of the flour is not important.

3.4.3 Soft Flours

Several sorts of flour are normally made from soft wheat. Some are high value products made to tight specification while others are not.

3.4.3.1 Plain Flour. This product is the ordinary flour used for most home baking. The equivalent product in the USA is general purpose flour, although one will not necessarily substitute for the other. This type of flour is sometimes called household flour.

British plain flour is made to a very wide specification that guarantees that it is white, unbleached and untreated. The Hagberg Falling number should not be too low.

A common grist would be mainly soft wheat with possibly some hard wheat added at up to 20% to encourage flowability and good mixing. Hard wheat that has failed a quality control for bead making can end up in plain flour.

While plain flour is intended as a domestic product it does find its way into some bakeries for some uses.

3.4.3.2 Self Raising Flour. This product is a soft wheat flour with a chemical raising agent, also known as leavening agent, added. It can always be substituted by a mixture of plain flour and baking powder.

An important property in self raising flour is that the moisture content should be kept below 13.5% lest the aerating reaction occur in the flour.

British self raising flour normally contains E500 sodium hydrogen carbonate (colloquially bicarbonate of soda) ($NaHCO_3$) and acid calcium phosphate, (ACP, E341 calcium tetrahydrogen diorthophosphate) [$CaH_4(PO_4)_2$]. A typical dosing rate would be 1.16% of the alkali and 1.61% of 80% grade ACP, all calculated on the weight of the flour.

An excess of the acid over the alkaline component is necessary as, if the flour is alkaline, on cooking an unpleasant odour and a yellow colour will appear.

In the UK, the following are permitted in self raising flour:

E500 sodium hydrogen carbonate (colloquially bicarbonate of soda) ($NaHCO_3$).
E341 calcium tetrahydrogen diorthophosphate [$CaH_4(PO_4)_2$].
E450 disodium dihydrogen diphosphate (sodium acid pyrophosphate) (SAPP).
E541 acidic sodium aluminium phosphate (SAP).
E570 D-glucono-1,5-lactone.
E336 mono potassium-L-(+)-tartrate (cream of tartar).

When ACP is used at 1.61% of the flour weight about 250 mg of calcium is added to each 100 g of flour, so there is no need to fortify the flour with calcium.

It is common to mix SAPP with starch to aid the addition. This mixture is known as "cream powder". When this is mixed in the ratio 2:1 with sodium hydrogen carbonate the combined product can then be added to the flour at 4.7% on the weight of the flour.

3.4.3.3 Cake Flour. This is also known as high ratio flour and was made by treating flour with chlorine gas. Originally, the chlorine was used to bleach the flour but it was found that the flour could be used to make cakes where the ratio of sugar to flour and of liquid to flour both exceeded one. Hence the expression "high ratio".

Heat treatment is now used instead of chlorination. The effect of either treatment is to reduce the elasticity of the gluten, presumably by denaturing the proteins. High ratio flour is particularly suitable for sponge cakes.

3.4.3.4 Biscuit Flour. Apart from yeast raised crackers, biscuits are made from soft wheat flour. The requirements of a good biscuit flour are almost opposite to those of a good bread flour. Biscuit flours are made from low protein soft wheat and should have a low starch damage. A high Hagberg Falling Number is no particular advantage, but a low Hagberg Falling Number can be an advantage. The dough characteristic required for biscuits is elasticity not resistance. Semi-sweet biscuits in particular need this sort of flour. Short dough biscuits are more tolerant of the flour used since gluten development is avoided. The only flour treatment of biscuit flours is with reducing agents, *e.g.* sulfur dioxide. This is usually achieved by adding sodium metabisulfite in the bakery. Biscuit flours are often above 70% extraction, which makes them brown rather than white flours. The higher extraction increases the yield in the mill and, hopefully, reduces the cost of the flour. Whiteness is not a highly regarded property in biscuits, which are expected to look slightly brown.

Some bread flour mills have difficulty in making a low starch damage soft wheat flour, a job that the mill was not intended for. This is probably why some millers do not make biscuit flours, leaving them as a niche product for the smaller milling companies.

3.4.3.5 Wafer Flour. Wafer flour is a type of biscuit flour with the same basic specification of low protein soft wheat flour with a low starch damage. Once again the required dough property is extensibility. The only differences are that if the protein is too low the wafer will be too soft to handle, and if the protein is too high the wafer will be too hard. The other important property is a resistance to gluten separation. Wafer flours are likely to be brown.

3.4.3.6 French Wheat Bread Flour. This again is a soft wheat flour with low starch damage but the wheat must be French bread making wheat. These wheats have the correct genetic make up to make bread. The protein content is higher then English soft wheat. Once again this is a product that British bread flour mills find it difficult to make. The problem being that they were designed to deliver a high starch damage to hard wheat.

3.4.4 Wholemeal Flour

Almost any type of flour can be made as a wholemeal in principle. Wholemeal flour is what it says it is, *i.e.* all of the wheat. It is illegal to bleach fortify or add flour treatments to wholemeal flour. It is the only type of flour that does not have to be fortified as the natural minerals that were present in the wheat are all present in wholemeal flour. The wheat germ oil is also present.

While any type of flour can be made in a wholemeal form some products probably can not be made from wholemeal flour. The problems are that although all the wheat protein is present the quality of the protein is lower towards the outside of the wheat berry. This effect would render the making of products requiring very high extensibility, *e.g.* filo pastry, very difficult. The other problem is that the bran particles tend to burst gas bubbles, reducing the amount of lift. Despite the above, wholemeal bread flour is common and wholemeal self raising flour can be made.

Bakers tend not to like wholemeal bread flour because it is less reliable in performance than white flour. One reason for this is the restrictive legal position on flour improvers, which makes the dough less tolerant. In addition, when a wholemeal flour is made in a roller mill, all the flour components that the mill has separated have to be recombined at the bottom of the mill. If a spout blocks temporarily one component will be held back. If this sort of problem leads to an excess of white flour then the flour will bake well, if a local excess of bran occurs then baking performance will suffer.

Although the 100% extraction rate ensures a bigger yield, wholemeal is not particularly popular with millers as a stronger grist is needed. In addition, its shelf life is only three months (cf. a year for white flour). The reduction in shelf life is supposed to be caused by oxidation of the lipid fraction that is absent from white flour. Furthermore, wholemeal flour must be kept apart from white flour less that is contaminated.

Claims are made about the health benefits of eating wholemeal bread. These claims are in regard to gastrointestinal health and cardiovascular disease. There does seem to be some truth in these claims.

3.4.5 Brown Flour

Brown flour is a term that covers the extraction rates above 70% and below 100%, *i.e.* between white and wholemeal flour. Unlike wholemeal flour the full range of flour improvers are legal. The sale of brown bread is lower than that of wholemeal. Some nutritional thinking points to brown flour as having most of the health benefits of wholemeal. Another advantage of brown flour is that the composition is controllable, unlike wholemeal which has to be the wheat as it comes.

Brown flour has all the shelf life and contamination problems of wholemeal flour. Some biscuit and wafer flours are brown merely because they do not need to be white.

3.4.6 Low Moisture Flour

The bread making properties of ordinary white bread flour improve for some time after manufacture, which is the origin of oxidative flour treatments. The shelf life of such a flour is around a year. If the moisture content can be kept low, either by selecting low moisture wheat or by drying wheat, the shelf life can be increased to three years. This sort of flour is supplied for ship's stores and similar purposes. This is the modern alternative to ship's biscuits.

3.5 LEAVENING AGENTS

While it is possible to leaven products either with yeast or chemical leavening agents, ultimately the product is expanded by gases and vapours. In a yeast raised product the expansion is caused by carbon dioxide and ethanol while in a chemically raised product the carbon dioxide is produced chemically.

3.5.1 Air

In mixing any dough some air becomes incorporated. This air, if it remains until the product is baked, will contribute to its expansion in the oven. Another function of the air bubbles is to act as nuclei for the formation of other bubbles such as carbon dioxide.

3.5.2 Water or Steam

The conversion of water into steam or water vapour involves an enormous increase in volume. Not surprisingly this causes the product to increase in volume.

Products where this is the sole expansion system are extruded products. This sort of cooking is used for the production of breakfast cereals, savoury snacks and some crisp breads. Of these, only the crisp breads are really within the scope of this book.

3.5.2.1 Extrusion Cooking. In extrusion cooking the ingredients are fed into the extruder either individually or as a pre-made slurry. The ingredients would obviously vary, depending on the product required, but are likely to include flour and water.

In the extruder the flour and water will be subjected to intense agitation and will be heated to above 100°C under pressure. Extruders are constructed to work at elevated pressures, indeed the barrel of the extruder is constructed in the same way as a gun barrel. Under these conditions the starch will gelatinise but the water cannot boil because the high pressure elevates the boiling point.

Once the product is allowed to emerge from the extruder the water flashes into steam, expanding the product to a foam. Any protein and emulsifiers present help stabilise the bubbles, which set as the product cools. As the water flashed to steam the latent heat of the steam is lost to the product thus cooling it rapidly.

Extrusion cooking is an ideal system for a large company. The equipment is capital intensive while the products can often be made from relatively low cost ingredients. Setting up an extruder is complicated, so extruded products are relatively hard to copy.

Extrusion cooking is a special case because a water-containing product can be heated above 100°C without the water boiling off. Except in pressure cooking, regardless of the oven temperature the interior of a baked product can not rise above 100°C until all the water has been driven off.

With bread, although oven temperatures of around 240°C are used the centre of a loaf may not rise above 60°C.

3.5.2.2 Frying. In a fried product, such as a doughnut, the fat is above 100°C and any water on the surface will flash into steam, but water in the interior will not reach boiling point except in very small or thin products.

3.5.2.3 Water in Oven Cooked Products. In oven baked products the inside of the product is not going to reach boiling point, indeed one of the simplest ways of obtaining a well-controlled temperature is to rely on holding a liquid at its boiling point. However, in a baked product, as the temperature rises the vapour pressure of the water rises, causing water to be lost by evaporation.

In sponge cakes made without chemical leavening, expansion in the oven can only come from the water and the air present. Similarly, meringues expand in the oven as the water evaporates.

When bread expands in the oven the resulting expansion is known as oven spring. It has been calculated that water expansion was responsible for some 60% of the expansion.[1]

A system in which water expansion is vital is puff pastry. The water content of the butter or margarine used as a fat is in just the right place to flash into steam and expand the product.

3.5.3 Yeast

The original leavening agent was yeast. Baker's yeast is one of the fundamental ingredients for bread. It is a single cell fungus and the strain used for making bread is *Saccharomyces cerevisiae.* Yeast can work in two different ways, aerobically or anaerobically. When yeast is working anaerobically it produces alcohol and carbon dioxide as below:

$$C_6H_{12}O_6 \rightarrow 2C_2H_5OH + 2CO_2$$

When it is working aerobically the reaction is:

$$C_6H_{12}O_6 + 6O_2 \rightarrow 6H_2O + 6CO_2$$

In both cases the substrate shown is glucose. In the second case the yeast is respiring. In practice respiration will rapidly exhaust the supply of oxygen. A yeast manufacturer is intent on producing yeast cells, so they will blow air into the process to maximise the growth of yeast cells. In some rapid bakery processes such as the Chorleywood and spiral mixer it is felt that initially incorporating air either by applying air pressure in the Tweedy mixer or because the spiral mixer incorporates air it will encourage the proliferation of yeast cells. These processes depend on ascorbic acid. As there is also the issue that ascorbic acid needs to be oxidised to act as a flour treatment.

The production of alcohol is important because its evaporation provides a significant part of the expansion in the oven, known as oven spring. Stauffer quotes that when a mass balance calculation was made on an experimental loaf based on 100 g of flour in 173.5 g of dough the loaf expanded by 360 mL on baking.[2] Of this 360 mL, 109 mL could be accounted for by carbon dioxide and air. The remaining 251 mL must be caused by the evaporation of alcohol and water. As alcohol forms an azeotrope (of approximately 95% ethanol and 5% water) that boils at 78°C it would be expected that all the alcohol would evaporate. On the basis that about 1% of the dough weight is ethanol this would be 1.7 g, which could vaporise to 830 mL, which is more than enough to account for the missing expansion.

3.5.3.1 Levels of Yeast Use. The level of yeast used varies between processes. Not too surprisingly the long bulk fermentation processes use the least yeast while the "no time" processes such as the Chorleywood and spiral mixer use the most. Thus, a bulk fermentation process could use yeast at 1% of the weight of flour while the "no time" processes would use 2–3% or more. Those products where the yeast has a hard time, such as rich dough products that are high in fat and sugar, tend to use high doses of yeast. High sugar levels or high levels of other soluble solids, *e.g.* polyols, put the yeast cells under high osmotic pressure, which makes fermentation difficult. English muffins tend to have a high dose of yeast used because mould inhibitors are employed and tend to inhibit the yeast.

3.5.3.2 Forms of Yeast. Yeast is available in several forms: compressed yeast, cream (effectively a liquid), dried into pellets and powders claimed to be instantly active. Bakeries normally use compressed yeast, which with cream yeast must be kept under refrigeration. A supply of dried yeast will always be kept handy lest the yeast delivery should fail or the refrigerator breaks down.

3.5.3.3 Dried Yeasts. The traditional form of dried yeast is known as active dry yeast (ADY). This product normally only had 75–80% of the gassing ability of a compressed yeast on an equivalent basis. ADY has to be rehydrated with water at around blood heat before it can be used.

More modern forms of dried yeast are now available, known as instant active dried yeast (IADY) and protected active dried yeast (PADY). These types of yeast can be mixed directly into dry ingredients, making them more effective at gas production than ADY.

Because the activity of dried yeast is reduced by exposure to oxygen, IADY is supplied vacuum packed or with an inert gas in the head space. PADY, which has the yeast encapsulated in fat, relies on an anti-oxidant for stability. PADY is essential for domestic bread machines.

3.5.3.4 High Sugar Yeasts. These are products specially produced to work under the high osmotic pressures in products like Danish pastries. These yeasts are available in the form of IADY products. There are also some Japanese strains of compressed yeast that can stand high osmotic pressures.

3.5.3.5 Yeast Enzymes. Yeast has a special relationship with enzymes as enzymes were first discovered in yeast. The name enzyme is derived from the German for "in yeast" (*i.e.* Zym). It was found that an extract produced from the yeast could undertake a fermentation. This at the time was a philosophically important point since it showed that there was nothing special about living organisms. We now know that enzymes are biological catalysts that, unlike like the sort of catalysts used in the chemical industry, can work at ambient temperatures.

Old text books refer to the enzyme zymase as being present in yeast. It is now known that zymase is a complex of fourteen enzymes.

3.5.3.6 Yeast Fermenting in Dough. When yeast is in a bread dough the traces of sugars present can be fermented directly. As yeast contains the enzyme invertase, any sucrose present can be inverted into dextrose and fructose which can then be fermented. If any dextrose from a high DE glucose syrup is present then it can be directly fermented. If there is any lactose present it can not be fermented at all. Similarly, any polyols such as sorbitol can not be fermented.

Once any directly fermentable sugars have been used up the yeast can only be fed by sugars produced from the starch. The only starch that can be broken down is the damaged starch. This explains why a certain amount of starch damage and some amylase activity is desirable in a bread flour.

The damaged starch is broken down by α-amylase and β-amylase in the flour. These produce maltose, which is split into dextrose by the maltase in the yeast. The yeast then ferments the dextrose to ethanol and carbon dioxide. If there is insufficient α-amylase present, *i.e.* the Hagberg Falling Number is too high, this reaction will only run slowly. When British bread flour was made from Canadian wheat this could happen and malt flour was added to the flour. Most millers now use fungal α-amylase instead. This has the advantage that it does not affect the Hagberg Falling Number and an overdose will not cause problems. An overdose of malt flour would render the dough impossibly sticky.

The reason for the non-response of the Hagberg Falling Number to fungal α-amylase is that it is inactivated at 75°C rather than the 87°C of cereal α-amylase. It turns out that fungal α-amylase preparations improve loaf volumes considerably. Most of this effect is produced by a lipoxygenase enzyme that is also present.

3.5.4 Chemical Leavening

3.5.4.1 Sodium Bicarbonate (Baking Soda). This material, variously known as bicarbonate of soda or baking soda, was the original chemical leavening agent. Baking soda can be used either on its own or combined with an acid (as in baking powder).

Sodium bicarbonate can only be used without acid in systems in which the baking soda reaches a high enough temperature (>120°C) to decompose thermally:

$$2NaHCO_3 + heat \rightarrow Na_2CO_3 + CO_2 + H_2O$$

The reaction with acid is:

$$NaHCO_3 + H^+ \rightarrow Na^+ + CO_2 + H_2O$$

Sodium bicarbonate is soluble in water at 0°C; a saturated solution is 6.5% with the solubility rising to 14.7% at 60°C. It can be expected then that sodium bicarbonate will dissolve in the aqueous phase of a batter or dough. It will then react with any acid present, including any acid ingredients such as butter milk. Chlorinated cake flour, where it is still used, has sufficient acidity (110 g of the flour will neutralise 0.27 g of sodium bicarbonate).

The rate of reaction of sodium bicarbonate with acid can be controlled by controlling the rate of dissolution of either the sodium bicarbonate or the acid. One way of doing this is by choosing different particle sizes for the sodium bicarbonate. Table 1 gives the specification for the different grades.

If a dry cake mix is being formulated for consumer use, the employment of grade 5 coarse granular grade would minimise loss of the active ingredient while the product is on the shelf. A common response when formulating this sort of product is to add more sodium bicarbonate than the recipe needs and hope that there is sufficient left when it is used. Using a grade of sodium bicarbonate with a large particle size reduces the need for this over use. Grade 1 powdered sodium bicarbonate decomposes on storage 50% faster than the coarse granular grade 5. Decomposition rates of 2–4% a week at 50°C, falling to 0.5–1% at 30°C, have been reported for the powdered grade.

3.5.4.2 Potassium Bicarbonate. Potassium bicarbonate has become available in commercial quantities for food use. The only reason for using it is that the sodium content of the resulting product is reduced. As the molecular weight of potassium bicarbonate is greater (100.11 for $KHCO_3$ compared with 84.01 for $NaHCO_3$) some 19% more is required to produce the same volume of carbon dioxide. Potassium bicarbonate is also more expensive. The reaction for its thermal decomposition is:

$$2KHCO_3 + heat \rightarrow K_2CO_3 + CO_2 + H_2O$$

3.5.4.3 Ammonium Bicarbonate. Ammonium bicarbonate is more popular in the USA than in the UK. It is often used as a supplementary leavening, particularly in biscuits and crackers. While it is stable

Table 1 *Specification for the different grades of sodium bicarbonate*

Grade Number	*minimum – maximum % cumulative retained*	*minimum – maximum % cumulative retained*	*minimum – maximum % cumulative retained*	*minimum – maximum % cumulative retained*	*minimum – maximum % cumulative retained*	*minimum – maximum % cumulative retained*	*minimum – maximum % cumulative retained*	*minimum – maximum % cumulative retained*
USS sieve:	60	70	80	100	140	170	200	325
Microns:	250	210	177	149	105	88	74	44
1. Powdered				0–2			20–45	60–100
2. Fine granular			0-trace	0–2			70–100	90–100
3. Fine powdered				0-trace	0–5		0–20	20–50
4. Granular			0-trace	0–2			80–100	93–100
5. Coarse granular	0–8	0–35		65–100		95–100		
6. Potassium bicarbonate	0–5			40–60			80–100	

dissolved at room temperature it decomposes at 40°C, *i.e.* in the early stages of the oven. The reaction for the decomposition is:

$$NH_4CO_3 \rightarrow NH_3 + CO_2 + H_2O$$

Thus, one mole of ammonium bicarbonate, *i.e.* 79 g, gives two moles of gas, *i.e.* 44.8 L.

Because ammonium bicarbonate can produce so much gas, precautions must be taken to ensure that it is uniformly distributed throughout the product lest large voids should appear in the finished item. The practical solution to this problem is to make a solution of ammonium bicarbonate in warm water and add that to the mixer rather than adding the solid with the other ingredients.

The ammonia produced by ammonium bicarbonate will expand the product successfully. However, as ammonia is water soluble, if the moisture content of the product exceeds around 5% the water will dissolve some of the ammonia, giving an ammonia taste.

There are exceptions to this rule as ammonium bicarbonate is successfully used in éclairs and other choux pastry products. It seems that the thin walls, large internal cavity and high baking temperature combined allow the ammonia to be driven off.

3.5.4.4 Acidulants. The other component of any system of chemical leavening based on sodium hydrogen carbonate is an acid. The original acidulants were sour milk (lactic acid), vinegar (acetic acid), lemon juice (citric acid) and cream of tartar (potassium acid tartrate). All of these will react immediately on mixing so that the carbon dioxide is released straight away. The product had to be baked before the carbon dioxide escaped from the batter or product. The only delay possible was that allowed by the batter viscosity.

The production of a baking powder depended on a system where the reaction could be delayed until required. In 1864 a patent was taken out to use monocalcium phosphate hydrate in making a baking powder. In 1885 sodium aluminium sulfate came to be used. This compound has a low solubility in water at ambient temperatures so it does not start to act until the product has heated up.

The combination of the two acidulants gave rise to the so-called double acting baking powder. After research up to the 1930s several substances were approved and available as acidulants.

Table 2 lists them and their reactions to produce H^+.

Some bakers buy baking powder others buy sodium bicarbonate and acidulants. Commercial baking powders are designed for different uses,

Table 2 *Acidulants and their reactions*

Acidulant	*Acronym*	*Reaction*
Monobasic calcium phosphate	MCP	$3Ca(H_2PO_4)_2 \rightarrow Ca_3(PO_4)_2 + 3HPO_4^{2-} + H_2PO_4^- + 7H^+$
Sodium acid pyrophosphate	SAPP	$Na_2H_2P_2O_7 \rightarrow 2Na^+ + P_2O_7^{2-} + 7H^+$
Sodium aluminium sulfate	SAIP	$NaH_{14}Al_3(PO_4)_{8.4}H_2O + 5H_2O \rightarrow 3Al(OH)_3 + Na^+ + 4H_2PO_4^- + 4HPO_4^{2-} + 11H^+$
Dimagnesium sulfate	SAS	$3Ca(H_2PO_4)_2 \rightarrow Ca_3(PO_4)_2 + 3HPO_4^{2-} + H_2PO_4^- + 7H^+$
Dimagnesium phosphate	DMP	$3Ca(H_2PO_4)_2 \rightarrow Ca_3(PO_4)_2 + 3HPO_4^{2-} + H_2PO_4^- + 7H^+$
Dicalcium phosphate.2H$_2$O	DCP.Di	$3Ca(H_2PO_4)_{2.2}H_2O \rightarrow Ca_3(PO_4)_2 + 3HPO_4^{2-} + H_2PO_4^- + 7H^+$

depending on whether a fast, slow or sustained release of carbon dioxide is required.

Monocalcium phosphate monohydrate reacts almost as quickly as cream of tartar (potassium acid tartrate). Anhydrous monocalcium phosphate has four-fifths of the reactivity. At ambient temperatures dicalcium phosphate dihydrate, sodium aluminium phosphate and some grades of sodium acid pyrophosphate are essentially unreactive.

At baking temperatures the other acidulants start to dissolve so that sodium aluminium phosphate reacts midway through the baking cycle while dicalcium phosphate is insoluble until 80°C but then triggers a late release of carbon dioxide, which prevents dips in the middle of cakes or collapses.

The reaction rate of sodium acid pyrophosphate can be controlled by adding calcium ions. While this process happens anyway in the presence of skim milk solids, manufacturers deliberately add calcium ions to sodium acid pyrophosphate to give grades with slower dissolution.

3.5.4.5 Baking Powders. The use of a manufactured baking powder is the commonest solution in bakeries. A single acting powder would use monocalcium phosphate anhydrous while household double acting baking powder would use monocalcium phosphate hydrate and sodium aluminium sulfate. Double acting baking powder intended for bakery use would use monocalcium phosphate hydrate and sodium acid pyrophosphate. The advantage of using sodium acid pyrophosphate is that different grades are available that allow the powder to be tailored to a particular use.

3.6 FLOUR TREATMENTS

3.6.1 Introduction

This section covers all the additives and treatments that are added or applied to flour. Some of these qualify as permitted flour treatments in law. The term flour improver is also used as a synonym. In this work the term improver is restricted to the compound improvers that are added to bread doughs. These mixtures tend to contain not only flour treatments but other required ingredients as well such as emulsifiers. The statutory additions that are made to flour for nutritional reasons are excluded. Also excluded are some of the substances that have historically been used but have now been universally banned. Some substances such as potassium bromate that are banned in the UK but are still legal elsewhere are covered.

3.6.2 Wholemeal Flour

The situation with wholemeal flour is refreshingly simple. Flour treatments are banned and there are no statuary additions. The addition of ascorbic acid to wholemeal flour is forbidden but the use of ascorbic acid in wholemeal bread is allowed. Presumably, it was thought beneficial to allow the change so that the Chorleywood plants could make wholemeal bread. The ascorbic acid presumably goes in as an improver with other ingredients.

3.6.3 Bleaching

Flour contains pigments that are mainly xanthophylls. If a suitable oxidising agent is applied the colour is bleached and a whiter flour and hence loaf is obtained. The advantage to the miller is that a whiter flour can be obtained at a higher extraction rate, thus increasing profitability. The practice is now banned in the UK. The use of benzoyl peroxide $(C_6H_5CO)_2O_2$ as a bleaching agent is still permitted in the USA. It has to be supplied mixed with inert inorganic fillers to prevent the risk of explosion. The fillers usually employed are $CaHPO_4$, $Ca_3(PO_4)_2$, sodium aluminium sulfate or chalk.

3.6.4 Oxidative Improvers

Long ago it was noticed that the baking quality of white flour improved with storage for 1–2 months. This effect occurred more rapidly if the flour was exposed to the air. During storage, initially the level of free fatty acids increases, presumably owing to lipolytic activity. Lipoxygenase activity then produces oxidised fatty acids as the proportion of linoleic and linolenic acids falls while the number of –S–S– bonds decreases.

Aged flour handles better, with more tolerance in the dough, giving larger loaves with a finer crumb structure. These beneficial changes can be accelerated by treating the flour with oxidising agents.

3.6.4.1 L-Ascorbic Acid E300. No chemist expects ascorbic acid to be listed as an oxidising agent as it normally behaves as a reducing agent. Ascorbic acid is even a permitted antioxidant! To act as a flour treatment ascorbic acid is oxidised to dehydroascorbic acid, which is a highly effective flour treatment but is itself unstable. The oxidation to dehydroascorbic acid involves the enzyme ascorbic acid oxidase. The action of dehydroascorbic acid on –SH groups is mediated by the enzyme dehydroascorbic acid reductase.

Ascorbic acid gives less increase in loaf volume than the same weight of potassium bromate and is more expensive. However, potassium bromate is now banned throughout the EU. Ascorbic acid is now permitted almost everywhere, with some countries such as Australia, Greece, Portugal and Germany not even bothering to limit the maximum level. Ascorbic acid appears to be entirely safe.

Ascorbic acid has always been an essential part of the Chorleywood process, but since potassium bromate has been banned higher levels of ascorbic acid are needed. The need to either treat the mixer with an oxygen enhanced atmosphere or to apply air under pressure in modern versions of the Chorleywood process is explained by the need to oxidise the ascorbic acid to dehydroascorbic acid. As oxygen is more soluble in water than nitrogen, increasing the pressure preferentially increases the oxygen level in the dough. This system makes ascorbic acid as effective as the combination of ascorbic acid and potassium bromate used previously. Fortunately for bakers it is possible to retrofit old Tweedy mixers for the new way of working (Chapter 6).

The action of ascorbic acid in the dough is to strengthen the gluten, which improves gas retention and hence the volume of loaves is improved. Ascorbic acid is faster acting than potassium bromate, which why ascorbic acid has always been associated with rapid processes.

3.6.4.2 Azodicarbonamide (1,1′-azobisformamide; $NH_2CONNCONH_2$; ADA). This oxidising treatment is normally supplied dispersed on calcium sulfate or magnesium carbonate to avoid the risk of explosions. The trade names Maturox and Genitron are used.

Azodicarbonamide works by oxidising sulfhydryl (–SH) groups, thus providing a dough improvement effect. Azodicarbonamide reacts very rapidly, typically being complete after a dough has mixed for 2–3 min. This speed of action precludes it being a substitute for potassium bromate, which is very slow acting.

Azodicarbonamide is legal in the UK, Canada, New Zealand, and the USA but not Australia or the EU (except the UK). As always the legal position should be checked. At the time of writing, azodicarbonamide appears to have escaped any health worries. The residue produced by its action is biurea, which does not appear to be a problem.

3.6.4.3 Chlorine Dioxide. Chlorine dioxide gas (ClO_2) is known as dyox. As a gaseous treatment it is normally applied at the flour mill. Dyox is widely used in the UK, USA, Australia and Canada and Japan.

The chlorine dioxide is made *in situ* by passing chlorine gas through an aqueous solution of sodium chlorite. Air is then passed through the solution to obtain a gas mixture with 4% chlorine dioxide that is applied to the flour at 12–24 mg kg^{-1}.

As well as acting as a flour improver, chlorine dioxide also bleaches the flour. Unfortunately it also destroys the tocopherols.

3.6.4.4 Potassium Bromate. Potassium bromate is now banned throughout the EU, including the UK. It remains in use in the USA but there is an agreement to minimise usage. It is still utilised in Canada and New Zealand but has been voluntarily discontinued in Japan. It is needed to operate the continuous mix systems employed in the USA. There is little doubt that potassium bromate is on the way out. Consumer pressure in the USA may force its withdrawal before legislation does. Section 3.6.8 below covers the health problems of potassium bromate.

In its action potassium bromate resembles the effect of ageing flour more than any other oxidiser. It has proved impossible to find a legal slow acting substitute for potassium bromate. This has caused bakers to discontinue using the ADD process (Chapter 7).

As with the other oxidising agents potassium bromate must be diluted on an inert filler such as calcium carbonate or calcium sulfate. Otherwise there would be a risk of an explosion.

Bromate treatment in the dough gives an increase in elasticity and a reduction in extensibility. These are of course the desired characteristics for making bread.

3.6.5 Reducing Agents

3.6.5.1 L-Cysteine. L-Cysteine has the unusual distinction of being permitted in both bread and biscuits. Although chemically it acts as a reducing agent in both systems it was part of the ADD method of making bread (Chapter 7). This process required a rapid acting reducing agent and a slow acting oxidising agent. Unfortunately, since potassium bromate has been struck off there is no suitable legal slow acting oxidising agent available. L-Cysteine remains legal in bread in the UK and many other countries. Although it is a naturally occurring amino acid it would still need approval for use in bread.

The permitted use level in bread flour in the UK is 75 mg kg^{-1} in all bread flour except wholemeal and biscuit flour. The use in biscuit flours is permitted at 300 mg kg^{-1}, except where sulfur dioxide or sodium metabisulfite is used. L-Cysteine is also used in pastry as a pastry relaxant. In both pastry and biscuits, not too surprisingly as the chemical action is the opposite of that in bread improvers, the reducing agents cause the extensibility of the dough to rise and the resistance to extension to decrease.

L-Cysteine is normally added as L-cysteine hydrochloride or L-cysteine hydrochloride monohydrate. When the ADD process was in use it was incorporated in a compound improver. If L-cysteine is used as a pastry relaxant it is supplied mixed with either soy flour or an inactivated wheat flour. This of course aids dispersion and the measurement of very small quantities.

3.6.5.2 Sulfur Dioxide and Sodium Metabisulfite. Sodium metabisulfite is merely used to generate sulfur dioxide *in situ*. Sulfur dioxide gas can be applied in the flour mill if required. The use of these substances in flour is only permitted in biscuit flours although both of them are permitted in several foods. Sulfur dioxide and sodium metabisulfite are not permitted if L-cysteine is present in flour.

Sodium metabisulfite can conveniently be added in the bakery as needed. When sodium metabisulfite is used as a pastry relaxant it is normally used as a compound improver mixed with either soy flour or heat treated wheat flour.

3.6.6 Cake Flours

3.6.6.1 Chlorine. Traditionally, cake flour was treated with chlorine gas at the flour mill. This produced the so-called "high ratio" cake flour, *i.e.* a flour that could be mixed with more than its own weight of both sugar and water. The use of chlorine is being phased out, to be replaced by heat treated flours. The heat treatment of flour does not need permission from anyone.

3.6.7 Sources of Enzymes

There are two sources of enzymes that are added to flour, neither of which count as improvers.

3.6.7.1 Fungal α-Amylase (FAA). This preparation was originally added to flour to supplement amylase levels in the same way that malt flour does. It has the advantage that fungal amylase stops working at 75°C while cereal amylase is not inactivated until 85°C. Thus fungal α-amylase does not affect the Hagberg Falling Number, nor will an overdose wreck a dough as an overdose of malt flour would.

It has since been discovered that the very considerable increase in loaf volume that fungal α-amylase preparations can produce is not caused by the amylase but by a lipoxygenase that is also present.

3.6.7.2 Malt Flour. Malt flour is the traditional source of extra α-amylase to add to flour. Now that British bread flour is substantially home grown there is little need of this supplementation. Some bakers would claim that malt flour improves the flavour of the bread. Malt flour is a food and not an additive.

3.6.8 Potassium Bromate Health and Legislation

Potassium bromate had been used as a flour treatment agent for over 80 years. It is a powerful oxidising agent and it became known that it is carcinogenic.

Initially, potassium bromate was assumed to react completely on baking. However, testing revealed that traces of it survived in to the finished product.

After 1990 the British government changed the law to remove potassium bromate from bread. This move was logical in the circumstances. It was convenient because most of the EU did not permit potassium bromate and its continued use in British flour was a bar to enter EU trade.

At the time of writing, the USA still permits the use of potassium bromate. It remains to been seen how much longer this use continues.

3.7 FOOD STARCH EXCLUDING FLOUR

Unmodified starch occurs as white or nearly white powders, as intact granules, and, if pregelatinised, as flakes, powders, or coarse particles. Food starches are extracted from any of several grain or root crops, including maize, sorghum, wheat, potato, tapioca, sago and arrowroot and hybrids of these crops such as waxy maize and high-amylose maize. They are chemically composed of either one or a mixture of two glucose polysaccharides (amylose and amylopectin), the composition and relative proportions of which are characteristic of the plant source. Food starches are generally produced by extraction from the plant source using wet milling sodium tripolyphosphate and sodium trimetaphosphate processes in which the starch is liberated by grinding aqueous slurries of the raw material. The extracted starch may be subjected to other non-chemical treatments such as purification, extraction, physical treatments, dehydration, heating, and minor pH adjustment during further processing steps. Wheat starch is a by-product of the production of dried vital wheat gluten. Food starch may be pregelatinised by heat treatment in the presence of water or by cold-water swelling.

Food starches are insoluble in alcohol, in ether, and in chloroform. If they are not treated to be pregelatinised or undergo cold-water swelling, then they are practically insoluble in cold water. Pregelatinised and cold-water swelling starches hydrate in cold water. When heated in water, the granules usually begin to swell at temperatures between 45 and 80°C, depending on the botanical origin of the starch. They gelatinise completely at higher temperatures.

Pure starch is not much used in bakery products. Some maize starch goes into a few types of biscuits or cakes under the heading of "corn flour". A few starch gels are applied on bakery flans.

3.8 FATS

These are sometimes referred to as oils and fats. The only difference between the two is that materials are normally referred to as fats if they are solids at room temperatures and oils if they are liquids. The terminology is not always consistent as the material from coconut is normally referred to as coconut oil when in fact it is a hard fat. Most animal fats are solid at room temperature while most vegetable oils are liquid at room temperature. However, there are animal fats that are liquid at ambient temperatures and vegetable fats that are solid.

Fats are used as ingredients in bakery products because they perform some function. As fats are generally more expensive than most other ingredients, commercial pressure would have eliminated them otherwise. In some cases the function of a fat can be either partially or completely replaced by some other ingredient, typically an emulsifier.

All fats are esters of fatty acids and the trihydric alcohol glycerol. They were known chemically as triglycerides but the preferred name is now triacylglycerols. The properties of a triglyceride depend on the fatty acids that are attached. The more carbon atoms that are attached the higher will be the melting point (Figure 1). If the fatty acids are unsaturated each double bond reduces the melting point.

Fatty acids can then be classified by their degree of unsaturation as well as their chain length. Thus, fatty acids are either saturated if they have no double bonds or unsaturated. Unsaturated fatty acids are further subdivided into monounsaturated diunsaturated and so on. Fatty acids with many double bonds are classed as polyunsaturated. Table 3 gives the approximate fatty acid compositions of several fats and oils, Table 4 lists the unsaturated fatty acids found in milk (note the very wide diversity).

When attempts were first made to substitute vegetable fats for animal fats the limited supply of hard vegetable fats, *e.g.* coconut, was a

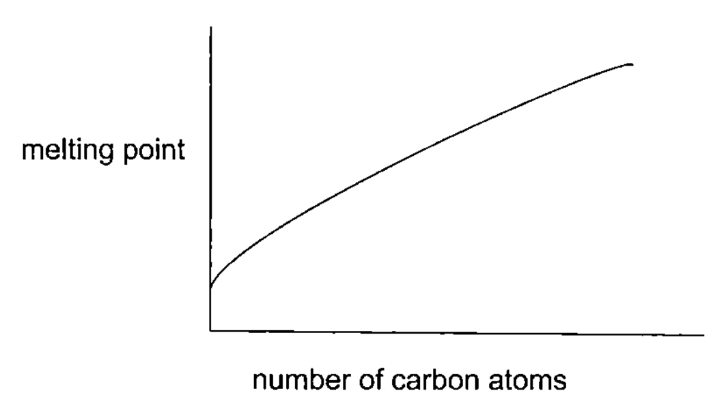

Figure 1 *Variation of fat melting points with number of carbon atoms*

problem. Many more vegetable fats were liquid at room temperature than were hard. Chemical investigation revealed that this softness was because the vegetable fats were more unsaturated than animal fats. If the double bonds were removed by adding hydrogen then a harder fat would be produced. The process of hydrogenating or hardening fats by treating them with hydrogen at high pressure over a nickel catalyst was a staple method of the oils and fats business for many years. However, catalytic hydrogenation yields products with trans double bonds rather than the cis conformation that exists in nature. Obviously, if the fat is completely hydrogenated there will be no double bonds and hence no problem; however, partially hydrogenated fats have trans double bonds. Trans double bonds are rare in naturally occurring fats, the major natural source is milk fat because they are formed by bacterial action in the rumen. It has emerged that fats with trans double bonds increase the risk of heart disease and so they are being phased out. Some countries have already placed restrictions on the level of trans fatty acids with advice that not more than 5% of energy intake should consist of trans fatty acids.

The loss of hydrogenation as a method is no doubt keenly felt. The effect on melting point of one double bond is about the same as having one less carbon atom. The oils and fats industry has re-organised and moved on to new technology. Some requirements are met by using fractionated fats while others are met by using interesterification. Fractionation is the process of breaking a fat into fractions by purely thermal means. This can be done by cooling the fat and removing the crystals or dissolving the fat in a solvent, *e.g.* acetone, and crystallising fractions from the solvent.

An alternative method is interesterification where the fatty acids are rearranged. This can be done chemically, which gives a random distribution, or by using enzymes. The advantage of enzymes is that they are very specific in their action. It is quite possible using a lipase to remove

Table 3 *Approximate fatty acid composition (%) of some fats and oils*

Source	C_4	C_6	C_8	C_{10}	C_{12}	C_{14}	C_{16}	C_{18}	$>C_{18}$	$<C_{16}$ *enoic*	C_{16} *enoic*	C_{18} *enoic*	$>C_{18}$ *enoic*	C_{18} *dienoic*	C_{18} *trienoic*
Butter fat	3–4	1–2	1–2	2–3	1–4	8–13	25–32	8–13	0.4–2	1–2	2–5	22–29	0.1–1	3	
Coconut				4–10	44–51	13–18	7–10	1–4				5–8	0–1	1–3	
Maize						0–2	8–10	1–4			1–2	30–50	0–2	34–56	
Cottonseed						0–3	17–23	1–3				23–44	0–1	34–55	
Olive oil					0–1	0–2	7–20	1–3	0–1		1–3	53–86	0–3	4–22	
Palm oil						1–6	32–47	1–6				40–52		2–11	
Palm kernel oil			2–4	3–7	45–52	14–19	6–9	1–3	1–2		0–1	10–18		1–2	
Peanut						0.5	6–11	3–6	5–10		1–2	39–66		17–38	
Soy bean						0.3	7–11	2–5	1–3		0–1	22–34		50–60	2–10

Table 4 *Unsaturated fatty acids present in milk*

Monounsaturated acids		*Polyunsaturated acids*	
Chain	*Percentage*	*Chain*	*Percentage*
10:1	0.27	18:2	2.11
12:1	0.14	18:2 cis,trans conjugated	0.63
14:1	0.76	18:2 trans,trans conjugated	0.09
15:1	0.07	20:2	0.05
16:1	1.79	22:2	0.01
17:1	0.27	–	–
18:1	29.6	18:3	0.5
19:1	0.06	18:3 conjugated	0.01
20:1	0.22	20:3	0.11
21:1	0.02	22:3	
22:1	0.03	–	–
23:1	0.03	20:4	0.14
24:1	0.01	22:4	0.05
–	–	20:5	0.04
–	–	22:5	0.06

the middle fatty acid. The resultant diglyceride could be used as an emulsifier or reacted with another fatty acid to produce a particular triglyceride. The oils and fats industry can then deliver whatever products are needed in terms of physical and chemical properties. Using technology it is possible to produce margarines that are easier to use in puffed pastry than butter.

3.8.1 Fat-containing Ingredients

Various fat-containing ingredients are in use in bakery products, which can be categorised as essentially pure fats, *e.g.* lard, largely fat ingredients, *e.g.* butter, ingredients with a substantial fat content, *e.g.* whole milk, and ingredients with traces of fat, *e.g.* wholemeal flour.

Ingredients with a trace of fat usually have little effect on the overall fat content but sometimes do alter the properties of the product. The traces of fat in materials often contain essential fatty acids.

The traces of fat in wholemeal flour reduce the shelf life of the flour as they become rancid fairly rapidly. Claims are made that this trace of fat has important dietary properties but the veracity or otherwise of this claim is beyond the scope of this book.

3.8.1.1 Hard Fats. These are substances like lard or hard vegetable fats such as hardened palm kernel oil (HPKO) or blended hard vegetable fats. These are normally weighed out and mixed into the system,

Table 5 *Specific requirements for lard specification*

	Rendered lard	*Bleached lard*	*Bleached-deodorized lard*
Colour (AOCS-Wesson)	Not more than 3.0 red	Not more than 1.5 red	Not more than 1.5 red
Free fatty acids (as oleic acid)	Not more than 1.0%	Not more than 1.0%	Not more than 0.1%
Hexane-insoluble water	Not more than 1.0%	Not more than 0.05%	Not more than 0.05%
Iodine value	Between 46 and 70	Between 46 and 70	Between 46 and 70
Unsaponifiable matter	Not more than 1.5%	Not more than 1.5%	Not more than 1.5%
Water	Not more than 0.5%	Not more than 0.1%	Not more than 0.1%

General requirements: *Identification*: Unhydrogenated lard exhibits the following composition profile of fatty acids, determined as directed under *Fatty Acid Composition*, Appendix VII:

Fatty acid	<14:0	14:0	14:1	15:0	16:0
Weight % (range)	<0.5	0.5–2.5	0.2	<0.1	20–32
Fatty acid	16:1	17:0	17:1	18:0	18:1
Weight % (range)	1.7–5	<1.0	<0.7	5.0–24	35–62
Fatty acid	18:2	18:3	20:0	20:1	
Weight % (range)	3.0–16	<2.0	<1.0	<1.0	

Lead: Not more than 0.1 mg kg^{-1}.
Peroxide value: Not more than 10 meq kg^{-1}.

possibly being melted first or possibly not. Table 5 gives the specific requirements for lard specification.

3.8.1.2 Butter. Similarly, butter would be weighed out and added to the product. Butter can be replaced with butter oil (see Table 10 in Section 3.14.5), which is butter with its water content removed. Butter oil has been used as a way of supplying butter to the bakery industry from intervention stocks with a reduced chance of the butter being diverted for table use.

Unlike margarine the properties of butter are fixed and can not be modified by chemical action. The only way to adjust the melting point of butter is to fractionate the fats to produce fractions with different melting points.

Initially, the dairy industry was interested in making a product that would spread straight from the refrigerator to compete with soft margarine. The excess of the hard fraction would either have to be mixed into the rest of the butter or sold as an extra hard butter.

Consumers did not take to the idea of extra soft butter so the project might well have ceased after that. Curiously, a market was found for the hard fraction amongst continental pastry cooks, patisseries and plant bakers. In some markets puff pastry croissants, *millefeuilles*, Danish pastries and similar products are preferred if they are made from pure butter. The specialised puff pastry margarines that the oils and fats industry had produced are easier to work with than ordinary butter. The extra hard butter gave the marketing benefits of butter with the performance benefits of margarine. Unfortunately no premium market has yet been found for the soft fraction.

3.8.1.3 Margarines. Margarine originated as a substitute for butter. The big advantage of margarine is that as a manufactured product the properties can be tailored to suit a particular use. Various bakery margarines are manufactured to meet the technical requirements of particular uses.

3.8.1.4 Pumpable Shortening. Plant bakeries often use a pumpable shortening rather than a margarine. On a large scale the pumpable shortening can be metered and pumped into the product automatically without using any labour.

3.8.2 Emulsifiers in Bread

While emulsifiers are normally used to either disperse oil in water or water in oil, or possibly air in either, they have special uses in bread. The use of emulsifiers in bread is tightly regulated. In bread appropriate emulsifiers strengthen the dough against mechanical abuse during processing and reduce retrogradation. The reduction in retrogradation has the practical effect of reducing staling, hence the shelf life is increased.

Two types of these emulsifiers are calcium and sodium stearoyl lactylates (CSL, SSL) and diacetyl tartaric esters of mono and diglycerides (DATEM esters). The bread and flour regulations 1984 permit the use of SSL at up to 5 g kg^{-1} in all bread while DATEM esters are permitted in all bread without limit. Typical use levels are around 0.5% on flour weight. CSL and SSL have been permitted in the USA since 1961.

3.8.2.1 Stearoyl-2-lactylates. These compounds can be regarded as the salts of reaction products between lactic and stearic acid. The products are sold as dough improving and anti-staling agents with claims for improved gas retention, shorter proving times and increased loaf volume. Claims are made regarding increased tolerance in the dough mixing process.

3.9 EMULSIFIERS

Emulsifiers are a class of substances that help to form or stabilise an emulsion. Some natural products, particularly gums and proteins, act as an emulsifier. Natural products often escape being defined legally as emulsifiers even though they undoubtedly are emulsifiers in practice. Substances capable of acting as an emulsifier tend to have one part of the molecule that is best suited to oily surroundings, *i.e.* it is said to be lipophilic, while the other end of the molecule is best in an aqueous environment, *i.e.* it is hydrophilic. The two opposite terms, *i.e.* hydrophobic meaning water hating and lipophobic meaning fat hating, are also in use. Emulsifiers are classified by a system of HLB numbers which refer to the ratio of hydrophilic to lipophilic groups present. Molecules with both hydrophilic and lipophilic groups are referred to as amphiphilic. Emulsifiers whether natural or synthetic in origin tend to be amphiphilic. An amphiphilic molecule is likely to be in its lowest energy state in the interface between an oil and a water phase. The diagram of interfacial layers is a gross over simplification of the nature of the interface in a real system. Real food systems tend to have a complex mixture of ingredients.

3.9.1 Foams

The type of emulsion of interest in bakery is foams. These can be regarded as a dispersion of air.

3.9.1.1 Sources of Emulsifiers. Some emulsifiers, *e.g.* lecithin, are purely natural products while others are manufactured, usually from natural materials. Typical materials for manufactured emulsifiers are vegetable oils, *e.g.* soya bean oil or palm oil, and animal fats, *e.g.* lard or tallow and glycerol. Where required, some manufacturers can supply products with kosher or halal certificates. Other raw materials are organic acids such as fatty acids, citric acid, acetic acid, tartaric acid, in addition to sorbitol and propylene glycol. One property that often affects the performance of emulsifiers is the purity. A very pure emulsifier will perform very differently to the same major ingredient in a lower purity. This is particularly apparent with monoglycerides, which are available as distilled monoglycerides produced by molecular distillation or in less pure grades.

3.9.1.2 Legislation. Emulsifiers tend to attract the attention of food legislators. It is entirely reasonable that only those substances that are safe for food use are permitted. However, it is difficult to understand why permitted emulsifiers vary so much between countries. The EU is working towards rationalisation in this area.

3.9.1.3 Examples of Emulsifiers. Distilled Monoglycerides E 471 – These are a high purity monoglycerides prepared by molecular distillation.

3.9.2 Lecithin

Lecithin is a naturally occurring emulsifier. It is even believed by some to be health food on its own. The discovery of lecithin goes back to the nineteenth century. In 1811 L. N. Vaquelin reported the presence of organically bound phosphorus in fat-containing extracts from brain matter. In 1846 Gobley separated an orange coloured sticky substance from egg yolk. The substance was found to have excellent emulsifying properties. Gobley named the substance lecithin, a name derived from "lekithos" the Greek for egg yolk. Lecithin is a polar lipid. The definition of a polar lipid being that it is a lipid that is insoluble in acetone. There are a whole class of phospholipids. Phospholipids tend to be found in membranes in animals and in plants.

3.9.2.1 The Definition of Lecithin. Lecithin for use in food is defined as "A mixture of polar and neutral lipids with a polar lipid content of at least 60%." NB this is different from the scientific usage where lecithin is used as a trivial name for phosphatidylcholine.

3.9.2.2 Sources of Lecithin. The main commercial source of lecithin is the soy bean. Lecithins are also produced from sunflower, rapeseed, maize and in small quantities peanuts. It can be produced from egg yolk but this is not commercially competitive. In the future it might be possible to produce lecithins from microorganisms.

3.9.2.3 The Production of Soy Lecithin. The soy beans are cleaned, de-hulled and cracked, followed by rolling to thin flakes. These flakes are then treated with solvent. The resulting mixture is known as a miscella. After filtration the solvent is vacuum distilled away. The residue is a reddish yellow oil that contains some 2% of impure lecithin. The oil and lecithin mix is heated to 70–90°C and mixed with 1–4% of water. This causes the lecithin to swell and precipitate as a jelly like mass that is then removed by specially designed high speed separators. The separated product is a sludge containing approximately 12% soy bean oil 33% phospholipids and 55% water. This material is then treated in a thin film vacuum evaporator to remove almost all of the water. The resulting product has 60–70% polar lipids, 27–37% soy bean oil, 0.5–1.5% moisture and 0.5–2% impurities. This product is the ordinary soy lecithin of commerce. It is possible to remove the soy bean oil to produce a de-oiled lecithin. In confectionery use there would be little point in using a de-oiled product. Lecithins can be modified chemically but this would cause them to lose their natural status. Another way of modifying the properties is to fractionate the raw lecithin, yielding products that are richer in one of the components. The resulting products retain their natural status.

3.9.3 Sucrose Esters E473

These emulsifiers are prepared from sucrose and edible fatty acids. The primary hydroxyl groups of the sucrose are esterified by the fatty acid. In Figure 2, R is the alkyl group of the fatty acid. Fatty acids can be reacted with one, two or three primary hydroxyl groups to yield mono, di or triesters, respectively.

One advantage of sucrose esters is that they can be made with a wider range of HLB values than other emulsifiers (Figure 3). Chemically, the families of emulsifiers shown in Figure 3 are all esters. As an emulsifier

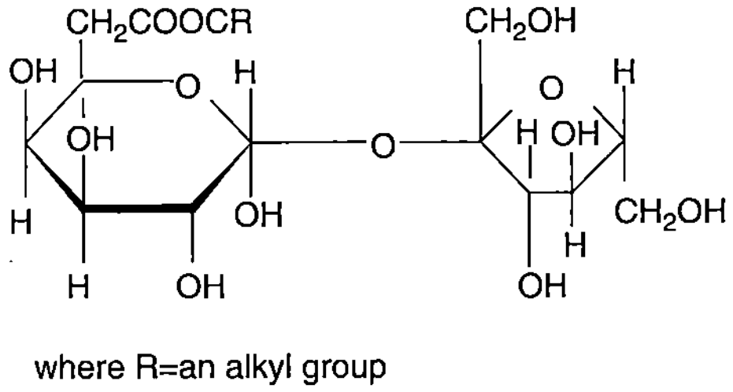

where R=an alkyl group

Figure 2 *Sucrose esters*

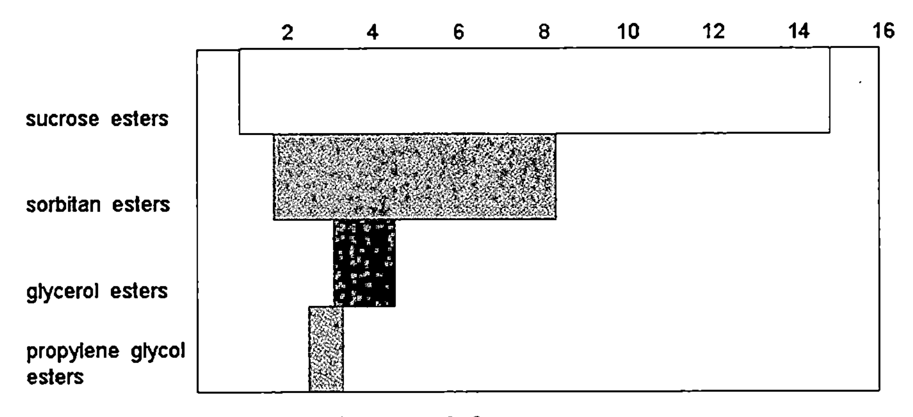

Figure 3 *HLB range of several ester emulsifiers*

Table 6 *Sucrose esters in confectionary products*

Product	*Use level(%)*
Soft chewy confectionery	0.1–2.0
Chewing gum	0.2–0.4
Tabletted products	0.1–2.0
Caramels and toffees	0.1–0.5

needs to be amphiphilic, esters are a popular structure for synthetic emulsifiers. Available grades of sucrose esters cover the HLB range 2–15. This wide range of HLB values is obtained by varying the monoester content from 10 to 70%. Thus a high HLB emulsifier would be suitable for use in an oil in water emulsion while a low HLB emulsion would be used in a water in oil emulsion.

The practical effect of this very wide HLB range is that sucrose esters can be used in a very wide range of confectionery products. Notably, it is not necessarily the same sucrose ester (Table 6). Sucrose esters are stable up to 180°C.

3.9.3.1 Regulatory Status. It might be thought that sucrose esters made from two food ingredients would have an easy passage in food legislation. This has not been the case. The early production method for sucrose esters involved the use of the solvent dimethylformamide (DMF). There were worries about the effect of DMF residues in the product. These problems have now been dealt with by tight specifications on residual DMF. Some sucrose ester production does not now involve DMF. The EU has included sucrose esters in directive 78/663 covering the emulsifiers permitted in all states of the EU. The E-number E473 has been assigned to all sucrose esters. The Scientific Committee for food has assigned an ADI of 20 mg per kilo of body weight. Sucrose esters are not permitted in bread in the UK.

3.9.4 Eggs

The most effective emulsifier and whipping agents are eggs and egg white. Of course these are ingredients rather than emulsifiers in law. Bakers would like not to use eggs for several reasons: they are expensive, do not keep well and are potentially contaminated with bacteria. The latter issue applies to any egg product that has not been heat treated. The main worry is that the raw egg will contaminate finished product.

Bakeries do not normally use fresh eggs but rely on either frozen whole egg or egg powder. The problem with egg powder is that it does not normally perform as well as less treated egg products.

While partial substitution of egg with emulsifiers is possible there is no viable complete replacement. The most promising egg replacers are enzyme-modified soy proteins.

3.9.5 Uses of Emulsifiers in Bakery Products

Emulsifiers are used in a range of bakery products. Their use in bread is covered in Section 3.8.2. In general, emulsifiers can substitute for some egg in sponge cakes. Emulsifiers soften the crumb of baked products and retard staling by impeding starch retrogradation.

3.9.5.1 Legal Restrictions. Bread is the most tightly restricted system because it has its own rules. Other bakery products fall under general food legislation. Of the emulsifiers lecithin is almost universally allowed while sucrose esters are the most tightly restricted. With monoglycerides the performance of the material is affected by the purity of the material. Distilled monoglycerides are an example of a high purity emulsifier.

3.10 COLOURS

When synthetic colours were first added to foods the dyes used were merely batches of the sort of dye used in the textile industry. Now of course food colours are rigorously tested to see that they are not harmful. The use of colours in foods is strictly regulated. Governments around the world have lists of permitted colours. Unfortunately, the lists differ throughout the world. It might be thought that some scientific consistency could be achieved but this is not the case. Indeed, some manufacturers who produce products for the international export markets are reduced to leaving out all the colours as a way of making the product universally acceptable.

Early fruit flavoured products were probably flavoured with jam and did not have a particularly strong flavour. Even with modern flavours the experiment of putting the "wrong" colour in the product will cause an appreciable proportion of tasters to misidentify the flavour.

3.10.1 Technical Requirements of Colours in Bakery Products

To be used successfully in bakery products a food colour needs the following attributes as well as complying with the appropriate legislation: it should be stable to heat and light, stable to reducing sugars, and resistance to sulfur dioxide is useful. Most colours used in bakery products are water soluble. This is simply convenience; some flour confectionery products contain very little fat any way.

3.10.2 Synthetic Colours

Synthetic colours are available for almost all possible shades. Intermediate shades can be produced by blending colours. In general, synthetic colours are much more stable than natural colours to light, heat and extremes of pH.

Synthetic colours can be supplied as soluble powders, prepared solutions, easily dispersed granules, pastes or gelatine sticks. Blocks of colour in vegetable fat are available for use in fat-based products. The attraction of soluble powders is that they are the least expensive and can be made up as required for use. The other forms have the advantage that they are at a concentration that is ready to use. The disadvantage is usually financial. Synthetic colours are normally so intense that they must be considerably diluted for them to be readily measured and dispersed into the product. Colour solutions made up in the factory have to be prepared not more than twenty four hours before use to avoid mould spoilage. The pre-prepared colour solutions will contain a permitted preservative or will be made up in glycerine, propylene glycol or propan-2-ol. These non-aqueous solvents inhibit mould growth. Table 7 lists some synthetic colours.

3.10.3 Natural Colours

There is a belief that natural products are inherently safer and more healthy than man-made ones. This belief is lacking in intellectual rigour. Of the most toxic substances known to man most are natural, *e.g.* aflatoxin, a mould metabolite, and ricin, found in castor oil beans. However, the presence of natural colours is a marketing advantage and so they are used. Natural colours in general are less heat stable, less light stable and give a less intense and less pure colour than synthetic colours. Natural colours have been used in the form of impure extracts rather than pure products. In this form higher doses are needed than with synthetic products. When purified, some natural pigments are more intense in colour and can be used in lower doses than synthetic colours. One other problem with natural colours is that the range of colours available is restricted. Several sources of natural colours are given in the following subsections (Table 8).

Table 7 *Some synthetic colours*

Shade	*Name*	*EU E No*	*USA FD&C No*	*Chemistry*
Red	Allura red AC	E129	Red 40	Monoazo
Red	Ponceau 4R	E124	Not permitted	Monoazo
Red	Carmoisine	E122	Not permitted	Monoazo
Red	Amaranth	E123	Not permitted	Monoazo
Red	Erythrosine BS	E127	Red 3	Xanthene
Red	Red 2G	E128	Not permitted	Monoazo
Orange or yellow	Tartrazine	E102	Yellow 5	Pyrazolone
Orange or yellow	Yellow 2G	E107	Not permitted	Monoazo
Orange or yellow	Sunset Yellow FCF	E110	Yellow 6	Monoazo
Orange or yellow	Quinoline Yellow	E104	Not permitted	Quinoline
Green	Green S (Brilliant Green BS)	E142	Not permitted	Triarylmethane
Green	Fast Green FCF	-	Green 3	Triarylmethane
Blue	Indigo Carmine	E132	Blue 2	Indigoid
Blue	Patent Blue V	E131	Not permitted	Triarylmethane
Blue	Brilliant Blue FCF	E133	Blue 1	Triarylmethane
Brown	Brown FK[a]	E154	Not permitted	Note 1
Brown	Chocolate Brown FB		Not permitted	Monoazo
Brown	Chocolate Brown HT	E155	Not permitted	Diazo
Black	Black PN	E151	Not permitted	Diazo

[a] Brown FK is a mixture of a monoazo and a diazo compound.

3.10.3.1 Caramel (E150). Caramel in this context means a brown colour that is produced either traditionally by heating sugar or as a very intense product that is made by heating carbohydrate, usually glucose syrup, with ammonia. Caramel colour is the product of the Maillard reaction, *i.e.* the reaction of a reducing sugar with an amino group. Chemically the colour is a melanoidin. These substances are extremely stable chemically and can be used in any type of confectionery.

3.10.3.2 Chlorophyll. This is the green pigment that is responsible for photosynthesis. It is widely distributed in nature, sources are green leaves, grass, alfalfa and nettles. The extract that is used is a mixture of chlorophyll with lutein and other carotenoids. This product gives an olive green colour. Chlorophyll is most stable in neutral or alkaline conditions but has a limited stability to heat and light. Chlorophyll preparations are available for colouring boiled sweets.

Table 8 *Natural colours*

Colour	*E number*	*Colour*	*Source*
Caramel	E150	Brown	Burning carbohydrates
Chlorophyll		Green	Green leaves
Copper chlorophyll	E141	Blue green	Chemically modified chlorophyll
Cochineal	E120	Red	Beetles
Riboflavin	E101	Orange yellow	Yeast or nature identical
Riboflavin-5-phosphate	E101a	Orange yellow	Nature identical
Carbon black		Black	Carbonised vegetable matter
Curcumin	E100	Yellow	Turmeric
Crocin		Yellow	Gardenia plants
Beta-carotene	E160(a)	Yellow or orange	Carrots or algae
Annatto	E160(b)	Orange	*Bixa orellana*
Lutein	E161(b)	Lemon yellow	Aztec marigolds or alfalfa
Betalaines	E162	Red	Beetroot
Anthocyanins		Purple red	Grape skins

3.10.3.3 Copper Chlorophyll (E141). This is made from chlorophyll; it is more blue than natural chlorophyll. The chemical modification makes it much more stable to heat and light. It is a more useful material than natural chlorophyll.

3.10.3.4 Cochineal (E120). Cochineal is a traditional natural colour. It is made from a Mexican beetle. The only problems with cochineal, apart from expense, is that it is not kosher and it is not animal free. Cochineal is not kosher not because it is made from an insect but because the insect is not itself kosher.

3.10.3.5 Riboflavin (E101). This is vitamin B2. Riboflavin can be extracted from yeast but is normally encountered as a nature identical substance. Unfortunately, riboflavin has an intensely bitter taste. The colour produced is an orange yellow. It is stable to acid but is unstable in water. Riboflavin is sometimes used for panned goods.

3.10.3.6 Riboflavin-5-phosphate (E101a). This material is both less bitter and more water stable than riboflavin. It is normally only encountered as a pure synthetic substance. Like riboflavin it is used on panned products.

3.10.3.7 Carbon Black. This is carbonised vegetable matter, *i.e.* very finely divided charcoal. Inevitably it is the most light fast of all colours. Obviously, it is only available as a solid. A common use is in liquorice products.

3.10.3.8 Curcumin (E100). Curcumin is obtained from the spice turmeric, which comes from the plant *Curcuma longa*, of the ginger family. Curcumin is obtained by extraction from the plant to give a deodorised product.

Curcumin is a bright yellow pigment that is oil soluble. It is sometimes produced in a water dispersible form.

The colour of curcumin varies with the pH of the medium. Under acid conditions a bright yellow is obtained but under alkaline conditions a reddish brown hue is obtained. This colour shift occurs because curcumin undergoes keto–enol tautomerism.

The most serious problem with curcumin is instability to light. One recommendation is that curcumin should not be used in products that are exposed to light unless the moisture content is very low. A confectionery product that fits this description is boiled sweets. The heat stability of curcumin is sufficiently good that it can withstand 140°C for 15 min in a boiled sweet mass.

The other stability problem with curcumin is sulfur dioxide. If the sulfur dioxide level is above 100 ppm then the colour will fade.

Within the restrictions outlined curcumin is a successful natural colour.

3.10.3.9 Carotenoids. The carotenoids are a wide range of substances. They are extremely abundant in nature. Natural production has been estimated to be 3.5 tonnes s^{-1}. Some 400 carotenoids have been identified to date. They are found in fruits, vegetables, eggs, poultry, shellfish and spices. Orange juice and peel contain 24 different carotenoids. Several carotenoids, *e.g.* β-carotene, are important as pro-vitamins. The ordinary diet contains large quantities of carotenoids, much greater than any quantity that might be used as a colouring agent. Dietary advice is to eat more carotenoids. β-Carotene is sometime erroneously referred to as vitamin A. In fact it is pro-vitamin A. The human body has a regulatory system that turns off the conversion of β-carotene into retinol (vitamin A) if stocks are adequate. This prevents hypervitaminosis. Overdoses of fat-soluble vitamins can be very serious since the body can not readily dispose of any excess. Thus, using carotenoids as food colours does not pose a risk of vitamin A overdoses.

Legally, carotenoids are divided between two E numbers. E160 covers the carotenoid hydrocarbons β-carotene, lycopene and paprika as well as the apo-carotenoids, *e.g.* bixin. E161 covers the xanthophylls and the carotenoids lutein, astaxanthin and canthaxanthin.

Most carotenoids are fat soluble, although preparations that allow them to be dispersed in water are made. The colours available from carotenoids vary between pale yellow and red.

Chemically carotenoids have conjugated double bonds that render them liable to oxidation. This tendency to oxidation can be diminished by adding antioxidants to the product. In the sort of product where natural colours are used suitable antioxidants would be tocopherols or ascorbic acid. Chemical antioxidants such butylated hydroxytoluene might be suitable technically but would not fit the image of an all-natural product. Ascorbic acid could be declared as vitamin C rather than as an antioxidant. Oxidation can be started by exposure to light and so this is best avoided. Carotenoids are generally stable to heat. The levels required can be as low as 10 ppm. β-Carotene is available as a nature identical form.

3.10.3.10 Crocin. Crocin is found in saffron and in gardenias. Extracting crocin from saffron is not economically viable. Saffron is obtained from *Crocus sativus.* Seventy thousand plants are needed to produce 500 g of saffron, which would contain 70 g of crocin. The commercial source of crocin is the gardenia bush. The town of Saffron Walden in Essex, UK, takes its name because saffron used to be produced there.

Chemically, crocin is the digentiobioside of crocetin. It is one of the few water-soluble carotenoids to produce a bright yellow shade in water. Unfortunately, crocin is bleached by sulfur dioxide levels above 50 ppm. The heat stability of crocin is good enough to use it in boiled sweets.

3.10.3.11 β-Carotene [E160(a)]. The natural sources that are exploited commercially for β-carotene are carrots and algae. The EU classifies β-carotene as E160(a). β-Carotene is an oil-soluble pigment, although forms that can be dispersed in water are available. The colour obtained varies between yellow and orange, depending on concentration. β-Carotene is stable to heat, sulfur dioxide and pH changes. It is, however, sensitive to oxidation, particularly when exposed to light. β-Carotene is successfully used to colour boiled sweets and other confectionery products.

3.10.3.12 Annatto [E160(b)]. Annatto is classified as E160(b). It is extracted from the seeds of a tree (*Bixa orellana*), which grows in America, India and East Africa. The extract is a mixture of two pigments, bixin and nor-bixin. Bixin is oil soluble while nor-bixin is water soluble. Both pigments are diapo-carotenoids. Annatto has long been used as a food colouring and has some uses as a food flavouring.

Both bixin and nor-bixin produce orange solutions. Bixin produces an orange solution in oily media while nor-bixin produces an orange aqueous solution. Obviously, bixin is the product of choice for high fat systems while nor-bixin is used in aqueous systems. Nor-bixin is one of the two water-soluble carotenes.

Nor-bixin is damaged by sulfur dioxide if the concentration exceeds 100 ppm. Acidic conditions or divalent cations, particularly calcium, can cause nor-bixin to precipitate. These problems are tackled by producing nor-bixin preparations with buffers and sequestrants.

Nor-bixin is relatively stable to heat. The most severe conditions will either isomerise the pigment or shorten the chain. Either of these changes will make the pigment more yellow. Nor-bixin can associate with protein, which stabilises the nor-bixin. The other effect of this association is to redden the colour.

3.10.3.13 Lutein [E161(b)]. Lutein is one of the four most common carotenoids found in nature. The EU classifies it as E161(b). Chemically, lutein is a xanthophyll and is similar to β-carotene. Although lutein occurs in all green leafy vegetation, egg yolks and in some flowers the commercial sources are the petals of the Aztec marigold and, to a lesser extent, alfalfa.

Purified alfalfa gives a clean, bright lemon yellow shade. Lutein is more stable to oxidation than the other carotenoids. It is also resistant to the action of sulfur dioxide. Lutein is oil soluble and is most effective dissolved in oil. Aqueous dispersible preparations based on lutein are available.

3.10.3.14 The Betalaines. The main pigment in the concentrated colour beet red is betanin. This is classified as E162 by the EU. The pure pigment is obtained by aqueous extraction of the red table beet. Approximately 80% of the pigment present in beetroot is betanin.

In an aqueous solution betanin gives a bright bluish red. The pure pigment is so intensely coloured that dose levels of a few parts per million are satisfactory. The problems with betanin relate to stability. Betanin is extremely sensitive to prolonged heat treatment. Short spells such as ultrahigh temperature (UHT) are tolerated. The conditions that make betanin unstable are oxygen, sulfur dioxide and high water activity. As confectionery is a low water activity system without sulfur dioxide or oxygen, betanin can be used.

3.10.3.15 Anthocyanins. Anthocyanins are water soluble and are responsible for the colour of most red fruits and berries. Some 200 individual anthocyanins have been identified. It has been estimated that consumption of anthocyanins is an average of 200 mg per day. This is several times greater than the average consumption of colouring material. There are claims made that consuming anthocyanins has health benefits.

Chemically, anthocyanins are glycosides of anthocyanidins and are based on a 2-phenylbenzopyrilium structure. The properties of the anthocyanins depend on the anthocyanidins from which they originate.

Anthocyanins are extracted commercially using either acidified water or alcohol. The extract is then vacuum evaporated to produce a commercial colour concentrate. The raw materials can be blackcurrants, hibiscus, elderberry, red cabbage or black grape skins. The most commonly used commercially are black grape skins, which can be obtained as a by-product.

Anthocyanins usually give a purple red colour. Anthocyanins are water soluble and amphoteric. There are four major pH dependent forms, the most important being the red flavylium cation and the blue quinodial base. At pHs up to 3.8 commercial anthocyanin colours are ruby red; as the pH becomes less acid the colour shifts to blue. The colour also becomes less intense and the anthocyanin becomes less stable. The usual recommendation is that anthocyanins should only be used where the pH of the product is below 4.2. As these colours would be considered for use in fruit flavoured confectionery this is not too much of a problem. Anthocyanins are sufficiently heat resistant that they do not have a problem in confectionery. Colour loss and browning would only be a problem if the product was held at elevated temperatures for a long while. Sulfur dioxide can bleach anthocyanins – the monomeric anthocyanins the most susceptible. Anthocyanins that are polymeric or condensed with other flavonoids are more resistant. The reaction with sulfur dioxide is reversible.

Anthocyanins can form complexes with metal ions such as tin, iron and aluminium. The formation of a complex, as expected, alters the colour, usually from red to blue. Complex formation can be minimised by adding a chelating agent such as citrate ions. Another problem with anthocyanins is the formation of complexes with proteins. This can lead to precipitation in extreme cases. This problem is normally minimised by careful selection of the anthocyanin.

3.11 FLAVOURS

Flavours are complex substances that can conveniently be divided into three groups: natural, nature identical and synthetic.

3.11.1 Natural Flavours

These can be the natural material itself; one example would be pieces of vanilla pod or an extract, *e.g.* vanilla extract. Extracts can be prepared in several ways. One is to distil or to steam distil the material of interest. Another is to extract the raw material with a solvent, *e.g.* ethyl alcohol. Alternatively, some materials are extracted by coating the leaves of a plant with cocoa butter and allowing the material of interest to migrate into the cocoa butter. These techniques are also used in preparing perfumery ingredients, indeed materials like orange oil are used in both flavours and perfumes.

In practice some natural flavours work very well; any problems are financial rather than technical. Examples of satisfactory natural flavours are any citrus fruit or vanilla. Some other flavours are never very satisfactory when all natural. Notably, citrus oils are prepared from the skin rather than the fruit.

3.11.2 The Image of Natural Products

The view exists that natural products are inherently safer and healthier than synthetic materials. Curiously, any new synthetic ingredient has to be most rigorously tested before it is allowed in foods. Natural products, provided their use is traditional, are normally allowed without testing. There is a legal distinction applied between an ingredient and an additive. In the UK, additives generally need approval while natural ingredients, provided their use is traditional, do not. Periodically, some natural substance is tested and found to have some previously unknown potential risk.

3.11.3 Nature Identical Flavourings

These are materials that are synthetic but are the same compound as is present in a natural flavouring material. From time to time it emerges that one substance produces a given flavour. Most chemists know that benzaldehyde has a smell of almonds. Some chemists know that hydrogen cyanide smells of bitter almonds. If a natural flavouring can be represented by a single substance and that substance can be synthesised then the flavour is likely to be available as a nature identical flavour. Vanilla flavour is a good example. Vanilla flavour can be all natural and derived from vanilla pods or nature identical or artificial. The nature identical product would be based on vanillin, which is in vanilla pods and has a flavour of vanilla. An artificial vanilla flavour would be ethyl vanillin, which is not present in vanilla pods but has a flavour two and a half times stronger on a weight basis than vanillin. The claim nature identical does not seem to be much appreciated in the English speaking countries. In some other countries it is an important claim for marketing purposes.

The qualification for nature identical varies between jurisdictions. In the EU, ethyl acetate made from fermented ethyl alcohol and fermented acetic acid is nature identical. In the USA, provided that the ethyl alcohol and acetic acid are natural, *i.e.* produced by fermentation, the ethyl acetate would be natural.

Practical flavours often contain a mixture of substances, some natural, some nature identical some synthetic. UK law classifies a flavour that contains any nature identical components as nature identical even though the rest of the flavour is natural. Similarly, the presence of any artificial components renders the flavour artificial.

3.11.3.1 The Case for Nature Identical Flavours. Although not much appreciated in English speaking countries, nature identical claims are more popular in German-speaking countries. Initially, it is difficult to see why a synthetic substance that happens to be present in a natural flavour should be preferred over a synthetic substance that is not present in a natural flavouring substance. Presumably, the advantage of a nature identical substance is assumed because it is thought to be inherently safe. This is a paradox since synthetic substances are normally tested for safety much more exhaustively than natural ones. Nature identical flavours do have the advantage over natural products that the price or quality is not affected by adverse harvests.

3.11.4 Synthetic Flavours

These are flavours that are produced synthetically but are not present in a natural flavouring material. The chemistry of flavours is a complex topic that has been the subject of many books, for example ref. 3. Synthetic flavours are made from a mix of flavouring substances that have been found to produce a given flavour "note". Those who develop flavours are referred to as flavourists. Flavourists take the musical analogy of notes further by referring to the top notes and the bottom notes of a flavour.

Flavour research is driven by a need to find compounds that produce desirable flavours. In some cases the improvement that is sought over the natural substance is not flavour intensity or cheapness but chemical stability.

One view of the way that flavours work is that they interact with certain receptors in the nose. Any other compound that has the same shape will work as well.

A typical synthetic flavour is a very complex mixture of substances. The mixture used will have been chosen to give the desired properties in the system of choice. Compounding flavours is a mixture of chemistry and sensory skills. Flavourists spend years learning how to produce flavours.

3.11.5 Dosing

Whether the flavour used is natural, nature identical, synthetic, or a mixture it has to be dosed into the product. Although some flavourings are very intense the volume added to the product has to be large enough for the equipment or the people to add it with sufficient accuracy. The flavour of course has to be uniformly distributed in the product. This normally means producing the flavour as a solution. Flavours are prepared for a particular use. As an example, citrus oil based flavours can be dissolved in various alcohols.

3.11.6 Developments in Flavours

The application of ever improving analytical methods will continue to reveal new flavouring compounds, be they natural, nature identical or synthetic. Not only are ever more sophisticated analytical techniques available but also improved methods of data analysis. The new science of chemometrics has developed to cope with the situation where chromatograms with hundreds of compounds are obtained.

Biotechnology could be applied to produce flavouring substances. If the gene responsible for producing a given substance can be identified then, in theory, that gene can be expressed in other organisms. No doubt the legislators will examine whether such products qualify as natural or nature identical and will come to several different conclusions. Conventional plant breeding methods are used to produce varieties of flavouring plants that give flavours with improved characteristics.

It remains an interesting speculation what would happen if a mutation of vanilla was produced that produced ethyl vanillin rather than vanillin. The new variety would be much more potent as a flavour. However, ethyl vanillin might then have to be classified as nature identical.

3.12 ANTIOXIDANTS

Antioxidants retard the oxidative rancidity of fats. The problem is caused by the addition of oxygen free radicals across any double bonds present. After rupture of the initial epoxide this leads to the production of various aldehydes and ketones. These can be very odiferous compounds. Oxidative rancidity comes on suddenly rather than gradually. One problem with oxidative rancidity is that it is a zero free-energy process and is not retarded by lowering the temperature. Antioxidants work by being a free radical trap, *i.e.* they readily combine with the oxygen free radicals, producing stable compounds. Any compound that has this ability is a potential antioxidant. Of course not all such compounds are suitable for use in foods. To be used a compound has to be non-toxic and must have legal approval for use. Some antioxidants are synthetic, a few are natural or nature identical.

3.12.1 Synthetic Antioxidants

The commonest synthetic antioxidants are butylated hydroxyanisole (BHA) and butylated hydroxytoluene (BHT). Other synthetic antioxidants are *n*-propyl gallate and *n*-octyl gallate. Any substance that can act as a radical trap will have antioxidant properties. There are strict rules governing the use of antioxidants in foods. Only those substances that are on the permitted list can be used.

3.12.2 Tocopherols

The major class of natural or nature identical antioxidants are the tocopherols. Tocopherols are naturally present in many plant tissues, particularly vegetable oils, nuts, fruits and vegetables. Wheat germ, maize, sunflower seed, rapeseed, soy bean oil, alfalfa and lettuce are all rich sources of tocopherols. Chemically the structure is a 6-chromanol ring with a phytol side-chain (Figure 4). α-, β-, δ- and γ-Tocopherols differ only in the number of methyl groups on the aromatic ring. At ambient temperatures the antioxidant activity is in the order $\alpha > \beta > \gamma > \delta$. At higher temperatures (50–100°C) the order inverts to give $\delta > \gamma > \beta > \alpha$. α-Tocopherol acetate is not an antioxidant since the active hydroxyl group is protected. The interest in this substance arises because under appropriate conditions, *e.g.* aqueous acidic systems, the tocopherol acetate slowly hydrolyses to give tocopherol.

Tocopherols are pale yellow viscous oily substances that are insoluble in water but are soluble in fats and oils. α-Tocopherol and its acetate are made synthetically. The synthetic products are racemates and are

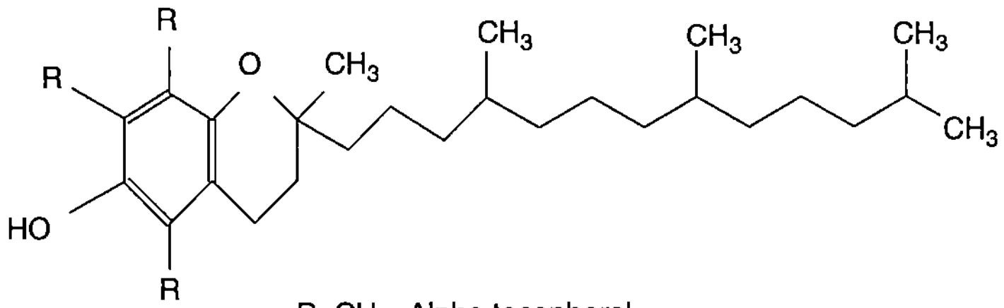

R=CH_3 Alpha tocopherol

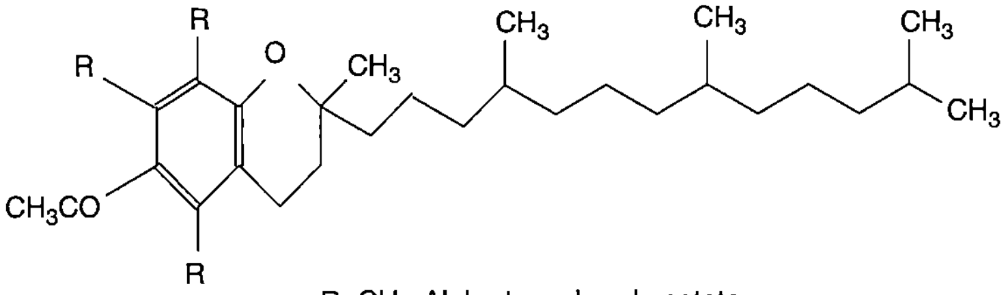

R=CH_3 Alpha tocopherol acetate

Figure 4 *Chemical structure of α-tocopherol and α-tocopherol acetate*

designated DL-α-tocopherol and similarly DL-α-tocopherol acetate. These are mixtures of the four racemates.

Tocopherols are not as effective as antioxidants as the synthetic antioxidants, *e.g.* BHA or BHT. The antioxidant effect of tocopherols is increased by mixing them with ascorbyl palmitate, ascorbic acid, lecithin or citric acid. Typical confectionery applications are the use of tocopherols with ascorbyl palmitate or lecithin or citric acid in the fat phase of toffees or caramels. Chewing gum base can be treated with α- and γ-tocopherol to extend the shelf life.

3.12.2.1 Natural Tocopherols (E306). The antioxidant E306 is defined as extracts of natural origin rich in tocopherols. This material can also be referred to as natural vitamin E. The major source of this material is the sludge produced by deodorising vegetable oils. The sludge will also contain sterols, free fatty acids and triglycerides. The tocopherols can be separated by several methods: one is to esterify them with a lower alcohol followed by washing, vacuum distillation and saponification. Alternatively, fractional liquid–liquid extraction is used. The product can then be further purified by molecular distillation, extraction or crystallisation. This process tends to produce a product high in γ- and δ-tocopherols. These can be converted into the more useful α-tocopherol by methylation. If required, α-tocopherol acetate can be made by acetylating the α-tocopherol.

3.12.2.2 α-Tocopherol (E307). This material is also known as synthetic α-tocopherol, synthetic vitamin E, and DL-α-tocopherol. This is the product formed by condensing 2,3,5-trimethylhydroquinone with phytol, isophytol or phytyl halogenides. The reaction is carried out in acetic acid or in a neutral solvent such as benzene with an acidic catalyst, *e.g.* zinc chloride or formic acid. The product is purified by vacuum distillation.

3.13 SUGARS

Sweet bakery products have developed around the properties of one ingredient, sucrose. The other bakery use of sucrose is to feed yeast.

Sucrose is a little unusual as a sugar since it is a non-reducing disaccharide. The constituent disaccharides are dextrose and fructose, both of which are reducing sugars. One of the crucial properties of sucrose is that its solubility at room temperature is limited to 66%. This means that a sucrose solution is not stable against bacteria or moulds. As an asymmetric molecule sucrose rotates the plane of polarised light. Now it is easily observable that if sucrose is heated with acid or alkali or treated with the enzyme invertase the optical rotation alters to the opposite direction. This is called inversion and has occurred because the sucrose has split into fructose and dextrose. The rate of the reaction can be measured by monitoring the optical rotation. In practice a small degree of inversion normally occurs when sucrose is boiled up in water.

Sucrose is extracted either from sugar beet or sugar cane. Normally, the two sources are equivalent even though the trace impurities are different. There is one area where the two sources are not equivalent and that is regarding brown sugars (Table 9). Cane sugar that has not been completely purified has a pleasant taste and can be used as an ingredient. Beet sugar is not acceptable unless it is completely white. In some products brown sugars or even molasses (the material left after sugar refining) are used to add colour and flavour. Alternatively, in some products a less than completely white product is used simply to save money. Beet sugar refiners do produce brown sugars. They are produced by adding cane sugar molasses to refined beet sugar. The brown sugars used in confectionery are carefully controlled products. They are not refined to high levels of purity but are produced with a carefully controlled level of impurity. Raw sugar is not normally used in bakery products. The only exceptions occur in health foods; very small tonnages of health food confectionery are made using raw sugar. Presumably the customers for this class of product believe that some benefit is conferred by using raw sugar.

Table 9 *Specifications for some brown sugars produced by British Sugar: these products are made by adding cane sugar molasses to white sugar produced from sugar beet; brown sugars can be made by partially refining cane sugar but not by partially refining beet sugar*

Type	*Light*	*Dark*	*Demerara*
Appearance	Fine, light golden brown	Fine, dark golden brown	Coarse, golden brown
Typical solution colour (ICUMSA units)	6.000	22.000	4.000
Typical particle size (μm)	300	300	700
Reducing sugars (min. %)	0.3	0.3	n/a
Loss on drying (max. %)	0.7	2.8	0.5
Total sugars (typical %)	99	96.4	99.4
Molasses addition (%)	2	8	1.3

Bakeries normally use sucrose in several forms: granulated, *i.e.* crystalline sugar, milled sugar, *i.e.* icing sugar, and possibly 66% sugar syrup. Sugar will normally be supplied to the factory in the granulated form. Sugar syrup is not stable and the economics of transporting large weights of water are not favourable; however, there are circumstances where it is easier to take in pre-dissolved 66% sugar syrup rather than making a sugar syrup on site. Supplying a mixture of sucrose and glucose syrup, which would be stable, is a better technical proposition.

Milled sugars have the problem that they are an explosive dust and must be handled with appropriate precautions. Some factories mill their own sugar on site while others have the sugar supplied pre-milled.

3.13.1 Molasses and Treacle

Molasses are the product left when no more sugar can be extracted. Beet sugar molasses are unpleasant in taste and are not normally used for human food. Cane sugar molasses do have some food use, normally in the form of treacle, which is clarified molasses. The ratio of sugar to invert sugar in treacle can be altered to some extent to assist product formulation. In practice different sugar syrups are blended with the molasses to give the desired product. Treacle is normally stored at 50°C to maintain liquidity.

3.13.2 Invert Sugar

Invert sugar is only encountered as a syrup. The fructose in the mixture will not crystallise so attempts to crystallise invert sugar yield dextrose. Invert sugar overcomes one of the big drawbacks of sucrose. Invert sugar solutions can be made at concentrations as high as 80%. These solutions have a sufficiently low water activity that they do not have biological stability problems. More importantly invert sugar can be mixed with sucrose and concentrated sufficiently to produce products that not only have a sufficiently low water activity to be stable but also will not crystallise. Adding invert sugar to a formulation lowers the water activity but makes the product hygroscopic. A few old fashioned products do not contain invert sugar as an ingredient but rely on the effect of heating sucrose in the presence of acid to generate some invert sugar *in situ*.

The use of invert sugar has declined since glucose syrup is cheaper and, for some uses, has superior properties. Some people take the view that invert syrup improves the flavour of some products.

There is another reason that encourages the use of invert sugar. Sugar-containing wastes can often be treated to produce invert sugar syrup. If a sugar solution is poured down a factory drain this generates a substantial charge for treating the resulting effluent. At the time of writing, a tonne of sugar costs £400. Allowing this tonne of sugar to become waste generates a cost of £200 per tonne. If the sugar can be recovered to produce invert, not only is the invert available as an ingredient that replaces some purchased material but the £200 per tonne disposal cost is avoided.

3.13.3 Glucose Syrup (Corn Syrup)

The ingredient known in the United Kingdom as glucose syrup has largely replaced invert sugar in many food uses. In the USA and some other English speaking countries this material is known as corn syrup. The syrup is made by hydrolysing starch to produce a mixture of sugars. Despite the name the major ingredient is not dextrose but maltose. In this work, to avoid confusion glucose will only be used to refer to the syrup while chemical glucose will always be referred to as dextrose. Originally the material was made by hydrolysing the starch with acid. This process was controlled by measuring the proportion of the syrup that gave a Fehling's titration and assuming it to be dextrose. Thus these syrups are specified in terms of "Dextrose equivalent", normally abbreviated to DE. Glucose syrup can be made from almost any source of carbohydrate. In practice it is only economic to produce it from maize starch, wheat starch or potato starch. Some wheat glucose is made as by-product of the production of dried wheat gluten. The process can be taken to completion to produce pure dextrose. This material obviously has a DE of 100. The commonest type of glucose syrup is 42 DE (or similar). This material is referred to as confectioner's glucose. Another common grade of glucose syrup that is used in bakery products is 68 DE, which has the same water activity as invert sugar syrup and so can often be used as a direct replacement for invert sugar syrup.

While glucose syrups were made by acid conversion the DE gave a complete specification of the product. The ready availability of suitable enzymes has widened the types of glucose syrups available enormously. Initially syrups became available that were produced by an acid plus enzyme process followed by products that were produced completely enzymically. The commercial advantage in this comes because a given weight of glucose syrup solids is cheaper than sucrose. The amount of sugar that can be replaced with glucose in a product is limited since 42 DE glucose is less sweet than sucrose and will affect the water activity and other properties. The glucose industry started to use enzyme technology to produce high maltose glucose syrups. These products had the same DE as confectioner's glucose but because there was a higher proportion of maltose in the product the sweetness was higher, allowing more sucrose to be replaced by glucose. The technology of the glucose industry has developed so far that virtually any starch hydrolysate could be produced if the demand was high enough.

The application of enzymes to glucose syrups was further extended to include the conversion of dextrose into fructose by glucose isomerase. The resulting syrups were known as "High Fructose Corn Syrup" in the USA or isoglucose or high fructose glucose syrup in the EU. The name comes because the product is produced normally from maize but using glucose isomerase. The initial product was a syrup that was chemically equivalent to invert sugar syrup. This product found a ready market in the soft drinks industry, particularly in the USA. In Europe, the authorities have not been keen on the idea of a product produced from starch that is possibly of non-EU origin replacing EU grown beet sugar. The conversion process can be continued to produce pure fructose. A product that has come into use in the UK is a glucose fructose syrup with 9.9% fructose.

3.13.4 Fructose

Fructose is normally encountered as a component of invert sugar. It has some properties that give rise to minor uses. Fructose is normally regarded as being twice as sweet as sucrose. High levels of fructose in a product tend to give a burning taste. One property of fructose that is sometimes useful is that, unlike other sugars, it is metabolised independently of insulin. For this reason fructose is sometimes used in products specially made for diabetics. It is claimed that small quantities of fructose smooth the taste of intense sweeteners when used in sucrose free products. Although fructose can be made from glucose syrup by using glucose isomerase, in Europe the common sources are chicory or Jerusalem artichokes.

Fructose is a very soluble and hence very hygroscopic product. It is usually used as a syrup. For many years fructose was referred to as the uncrystallisable sugar. Attempts to crystallise it by normal methods do not work. Fructose in a form that is described as crystalline is now available commercially. The product could well be produced by spray drying.

3.13.5 Dextrose

Pure dextrose is sometimes used as a food ingredient. It has roughly half the sweetness of sucrose. In Europe, the use of dextrose is not particularly attractive commercially; however, in other parts of the world it use can be economically advantageous.

3.13.6 Lactose

Lactose is a disaccharide reducing sugar. Unlike the other sugars mentioned, lactose is not particularly soluble. A property that has some use in yeast-containing products is that lactose is not fermented by baker's yeast.

Some individuals are unable to metabolise lactose and are lactose intolerant. This is because they lack the enzyme lactase that is needed to metabolise lactose. Lactose intolerance is common in those parts of the world where humans do not consume any dairy products after weaning. In practice this means in Asia, which means that most of the world's population might be lactose intolerant. It is possible to produce lactose removed skim milk. Another approach with lactose is to hydrolyse it to its constituent monosaccharides. As well as avoiding lactose intolerance this allows a syrup to be produced from cheese whey. These syrups are offered as an ingredient for toffees and caramels.

Lactose is normally encountered as a component of any skim milk that is used in bakery products. Small quantities of crystalline lactose are sometimes used in baked goods. If a product is made with too much lactose then a metallic taste appears. The amount of lactose that can be consumed without this taste appearing varies between individuals.

As one of the effects of the Common Agricultural Policy (CAP) has been to increase the price of all milk products there has been some substitution of skim milk powder by products derived from whey. Impure grades of spray dried lactose derived from whey are offered as a bakery ingredient. Adding lactose to these systems encourages the Maillard reaction, giving improved colour and taste.

The CAP has succeeded in reducing the amount of milk powder used in bakery products in Europe. The high price of skimmed milk encouraged manufacturers to look for substitutes and in some cases reduce levels of use that had started when milk powder was much cheaper.

3.14 DAIRY INGREDIENTS

Bakery products are not normally made directly from liquid milk although some home baking recipes do work by using liquid milk where water might otherwise be added. The problems of keeping and using fresh milk in a bakery are too great.

Milk solids are normally used as either milk powder or sweetened condensed milk in food manufacturing. Skim milk solids are an essential part of toffees as well as contributing useful colour and Maillard reaction flavours to baked goods.

Originally, full cream milk solids were used but now where possible skim milk solids are substituted. A few products are still made from full cream milk solids but this is now rare. In some cases butter or butter oil is added to replace the fat that has been removed from the skim milk. In other cases the fat content of the milk is replaced with vegetable fat. It might appear curious that whole milk is effectively reconstituted from skim milk and butter but there are good reasons. Skim milk powder keeps better than full cream milk powder. Using skim milk and butter can under certain conditions be economically advantageous.

In terms of performance in the product it is much easier to replace milk fat with vegetable fat, possibly adding a butter flavour, than to replace skim milk solids. Since milk fat has been the more expensive component for some years, financial pressures have encouraged the replacement of milk fat.

3.14.1 Sweetened Condensed Milk

The preferred source of milk solids for making toffee and caramel products remains sweetened condensed milk. This was one of the earliest ways of producing a stable product from milk. Both full cream and skimmed milk forms are used. The advantage of skimmed sweetened condensed milk is that the milk fat can be replaced with vegetable fat if so required. Products made from sweetened condensed milk are normally smoother than those made from milk powder. Presumably, the milk protein in sweetened condensed milk is in a less damaged form than in milk powder. Sweetened condensed milk has the advantage that provided the tin is not opened it keeps well without refrigeration. Sweetened condensed syrup is a sticky syrup and needs some skill in handling; however, those factories that use it are expert in using sticky syrups.

3.14.2 Evaporated Milk (Unsweetened Condensed Milk)

Evaporated milk is a more modern product than sweetened condensed milk. It is not normally used in food manufacture. This material has no technical or economic advantages over milk powder.

3.14.3 Milk Powder

Milk powder is the other form in which milk solids are used in bakery products. Both skim milk powder and whole milk powder are used. Skim milk was originally roller dried but this process has now almost passed out of use. Modern milk powders are made by spray drying. This does less damage to the proteins than the older roller process. The bioavailability of the proteins in spray-dried powder is higher than in roller dried powder. In bakery products this is not a particular advantage since these products do not form a major part of the diet. The less severe heat treatment of modern milk powder production can lead to problems since enzymes present in the milk are not inactivated. The enzyme in milk products, particularly milk powder, that causes problems is lipase. It should be appreciated that this is not the native lipase of milk but refers to bacterial lipases produced during storage. While the native lipase of milk is relatively easily deactivated, bacterial lipases are much more resistant to heat treatment. The bulk cold storage of milk does seem to favour organisms that produce heat resistant lipases. Lipase splits fatty acids from glycerol to produce free fatty acids. If the original fat was butterfat then at low levels this produces a "buttery" or "creamy" flavour. As the free fatty acid content increases the flavour shifts to cheesy. Normally, in baked goods free butyric acid is not a problem at any practical level, possibly because of losses during cooking. Other free fatty acids have different flavours. Lauric acid, which is found in nuts, tastes of soap. This is not too surprising as soap often contains sodium laurate. Lauric fats such as hardened palm kernel oil are often used as a substitute for butter. Another potential source of lauric fats is nuts that are sometimes incorporated in bakery products.

In any of these cases lipolytic activity can shorten the shelf life of the product or render it totally unacceptable.

3.14.4 Butter

Butter is the principal form of milk fat as an ingredient in baked products. The manufacture of butter is one of the two oldest dairy products, the other product being cheese.

Traditionally, butter was made by allowing cream to separate from the milk by standing the milk in shallow pans. The cream is then churned to produce a water in oil emulsion. Typically butter contains 15% of water. Butter is normally made either sweet cream or lactic, also known as cultured, and with or without added salt. Lactic butter is made by adding a culture, usually a mixture of *Streptococcus cremoris*, *S. diacetylactis* and *Betacoccus cremoris*. The culture produces lactic acid as well as various flavouring compounds, *e.g.* diacetyl, which is commonly present at around 3 ppm. As well as any flavour effect the lactic acid inhibits any undesirable microbiological activity in the aqueous phase of the butter. Sweet cream butter has no such culture added but 1.5 to 3% of salt is normally added. This inhibits microbiological problems by reducing the water activity of the aqueous phase. It is perfectly possible to make salted lactic butter or unsalted sweet cream butter if required. In the UK most butter is sweet cream while in continental Europe most butter is lactic.

Another type of butter is whey butter. This is produced from cream that has been skimmed off whey after cheese making. The cream in whey butter has been subjected to the controlled lactic fermentation used in cheese making. As a consequence whey butter has a characteristic and stronger flavour than other butters. Any type of butter can be used to make baked products. Traditionally toffee makers preferred to buy rancid butter if available. As butter is stored lipolysis causes the quantity of free fatty acids to rise. One of these fatty acids, butyric acid, at low levels gives a pleasant buttery flavour. At higher levels the flavour becomes cheesy and at still higher levels takes on notes of Parmesan cheese. The response to butyric acid varies between individuals. Some individuals would regard butter as improving with storage. Butter flavours tend to contain butyric acid. An approach that is used is to add a small quantity of lipolysed butter to the product. This has the same effect as using stale butter or adding a butter flavour. The lipolysed butter is a small quantity of butter that has been deliberately treated with a lipolytic enzyme to release the fatty acids. One advantage of lipolysed butter is that it can be described as "all natural" and, depending on legislation, is treated more kindly than a chemical flavour would be.

A common mistake about butter is the assumption that the fat must be crystalline because it appears to be a solid. In fact to completely crystallise all the fat in butter the butter must be stored at –40°C [cf. the normal temperature of a deep freeze at –20 to –18°C (0°F)]. Butter does not become completely liquid until 38–40°C. This is an extremely wide crystallisation and melting range. Unlike processed vegetable fat the composition of butter can only be altered by fractionation. It is possible to fractionate butter by several methods, using solvents such as acetone or alcohol, vacuum distillation, or slow crystallisation. Solvent fractionation can be used to produce well-defined fractions but has disadvantages. The solvent residues have to be removed but volatile aroma compounds tend to get lost in the process. The original interest in fractionating butter came about to produce a butter that would spread straight from the refrigerator, *i.e.* a butter to compete with soft margarine. Any fractionation process will produce more than one product so a use had to be found for the hard fraction. Hard fats make it easier to make puff pastry. In some countries pure butter puff pastry products, *e.g. millefeuilles*, are much appreciated. The hard fraction has turned out to be excellent for this purpose. Unfortunately, demand for the soft fraction has not been sufficient.

3.14.5 Butter Oil (Anhydrous Milk Fat)

Butter oil is covered by an international dairy federation specification for anhydrous milk fat (IDF standard 68a 1977) (Table 10). This product is milk fat with the water content reduced to 0.1% or less. It can be made by concentrating cream to 75% followed by treatment in a phase inverter and subsequent centrifugal separation. It is more common to make butter oil by melting the butter and removing the water with a centrifugal separator. At one time butter oil was being made from butter that had been held in intervention stores. Butter oil has a very long shelf life. It avoids the problems normally associated with storing butter. In some countries with no milk production butter oil is combined with skim milk powder to produce milk products such as sweetened condensed milk, evaporated milk, ice cream and UHT milk.

3.14.6 Whey

Whey is the by-product of cheese making. The traditional form of whey in the food industry is whey powder. This powder has been used as an

Table 10 *Specification for butter oil from IDF standard 68[a] 1977 for anhydrous milk fat*

Milk fat	99.8% minimum
Moisture	0.1% maximum
Free fatty acids	0.3% maximum expressed as oleic acid
Copper	0.05 ppm maximum
Peroxide value	Not greater than 0.02 (milli-equivalents of oxygen per kilo of fat)
Coliforms absent in 1 g	
Taste and odour	Clean, bland (samples to be between 20 and 25°C)
Neutralising substances	Absent

[a] This specification is for butter oil, which is butter with the water removed. The free fatty acid limit is to detect lipolytic rancidity while peroxide value specification is to limit oxidative rancidity. The copper limit arises because copper catalyses the oxidation of fats. The absence of neutralising substances is to prevent a high titration for free fatty acids being covered up by the addition of alkali.

ingredient in some bakery products. The major ingredient is lactose, which acts as another restriction on use. Lactose has a limited solubility compared with other sugars and when used to excess imparts an unpleasant metallic taste to the product.

New technology has been applied to allow more whey to be used. The technology has been applied to convert the lactose into a mixture of dextrose and galactose. These two monosaccharides are both reducing sugars and the mixture is much more soluble than lactose.

3.14.7 Vegetable Fats

Vegetable fats are mainly used in bakery products as a substitute for milk fat. This is particularly so in the EU where the CAP increases the price of milk fat. Because vegetable fats can be blended hydrogenated and interesterified it is possible to produce a vegetable fat with almost any desired range of properties. In general, the fats that are used as a substitute to milk fat are not an attempt to match the composition of milk fat but are designed to provide the best blend of properties for the product.

3.15 GUMS AND GELLING AGENTS OR HYDROCOLLOIDS

Another title that some of these ingredients fall under legally is thickeners and stabilisers. These ingredients are normally only minor components of bakery products and can properly be regarded as additives.

In bakery products they are used to produce glazes, jams and pie fillings, where the ability to act as a thickener and stabiliser is useful. The thickened filling can have a better texture with less tendency to make the pastry soggy.

Gelling agents under appropriate conditions self-associate to produce a three-dimensional structure. Some gelling, as with gelatine, is thermo-reversible while others, such as with high methoxyl pectin, is irreversible. Apart from the effects on the texture of the product an irreversible gelling agent is more of a problem in the factory since it cannot readily be recycled.

3.15.1 Agar Agar E406

The name is not a printer's error, although the substance is often referred to as agar. This gelling agent is a seaweed polysaccharide. It is extracted from red seaweeds from Japan, New Zealand, Denmark, Australia, South Africa and Spain although there are other possible sources. Initially the agar is extracted from the seaweed with hot water. The resulting solution must be concentrated – typical methods would be freezing and thawing or by heating under vacuum. The vacuum reduces the boiling point, which both saves energy and reduces damage to the agar. The colour of the finished product is improved by the use of bleach. The commercial form of agar can be either strips, flakes or powder. The finished product can have a characteristic flavour and odour. Typically the molecular weight is over 20 kDa.

The strength of gel given by agar is variable, depending on the origin of the agar. This makes it necessary to test batches of agar for gel strength before use. As the strength of gel is also dependent on the pH and the total solids the strength test needs to be under the conditions of use. If agar is used in fruit jellies the typical pH would be between 4.5 and 5.5.

In use agar is normally dissolved by mixing with ten times its own weight of sugar and dissolving in 30 to 50 times its own weight of water. The sugar is added to prevent the agar forming lumps on addition to water. This technique is often used when adding a slow dissolving solid to water as the sugar dissolves quickly, dispersing the agar. Alternatively agar can be added to boiling water. Temperatures above 90°C (184°F) are needed to dissolve agar. Agar is resistant to heat unless the conditions are acid. It is normal to reduce the temperature to 60°C before adding any acid. The maximum gel strength is obtained at pH 8 to 9 with the solids between 76 and 78%. If the solids are above 80% the gel strength is reduced. The gel strength is not directly related to the proportion of gelling agent used. In a typical confectionery fruit gel 0.5 to 1.5% of agar would be used. The texture of an agar jelly is also altered by adding jams or fruit pulps. The typical texture of an agar jelly is described as short. The effect of jam and fruit pulp is likely to be caused by the pectin naturally present. The gel strength of agar jellies is increased by adding locust bean gum but is reduced by adding alginates or starch. Agar is not considered to carry fruit flavours well. Agar is primarily used in bakery products in bakery icings and glazes.

Agar needs more water to get it into solution than do other gelling agents. Agar gels display hysteresis in melting at 85–90°C (175–184°F) but setting at 30–40°C (86–104°F). This is a useful property when the product is being deposited hot into moulds as there is no possibility of premature setting or pre-gelling. It takes longer for an agar gel to set in starch moulds than an equivalent product made from pectin. Agar is more difficult to handle in automated plants than some equivalent products. One advantage of agar is that it is not metabolised in the body. This does lead to a slight reduction in energy content.

3.15.2 Alginates E401

These materials are another polysaccharide. The name derives from its original source, brown algae. The current commercial sources are brown seaweeds such as *Laminaria digitata*, *L. hyperborea*, *Ascophyllium nodosum* and *Fucus serratus*. Different properties are obtained in alginates from different seaweeds. The sources are rocky coasts in the United States, the United Kingdom, France and Norway.

The alginates are extracted by first treating the seaweed with a dilute mineral acid. This converts the alginate into alginic acid. The alginic acid is insoluble, which allows any mannitol or mineral salts present to be washed away. The purified demineralised seaweed is then treated with alkali and ground. The alginic acid is thus converted into soluble alginates. Insoluble impurities such as cellulosic and proteinaceous materials can then be removed by filtration, flotation and settling. The alginates are then precipitated as alginic acid by adding acid. The precipitate is then washed and dried. The alginate required can then be made by adding the appropriate alkali. The finished product can then be milled and sieved to the appropriate size.

Calcium alginate has the formula $[(C_6H_7O_6)_2Ca]_n$. The molecular weight is in the range 32 000–250 000. It is insoluble in water, acids and organic solvents but is soluble in alkaline solutions. The monomers of alginate are mannuronic (M) and guluronic (G) acids (Figure 5). The alginates are composed of homogeneous M-M segments, homogeneous G-G segments and heterogeneous M-G-M-G segments. Of these segments only the guluronic acid segments can form the so-called egg box

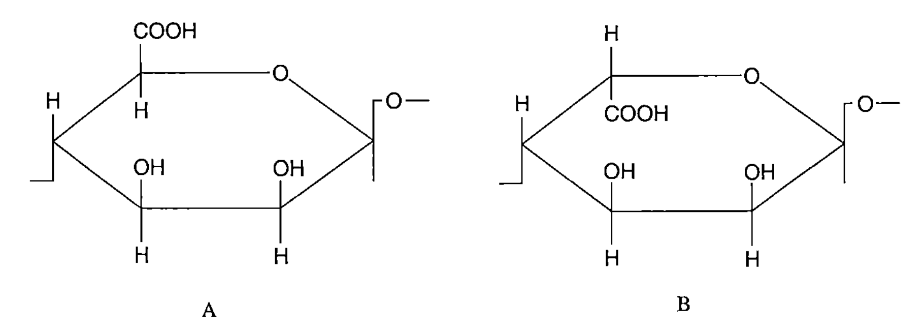

Figure 5 *Mannuronic (A) and gluronic acid (B) components of alginate monomers*

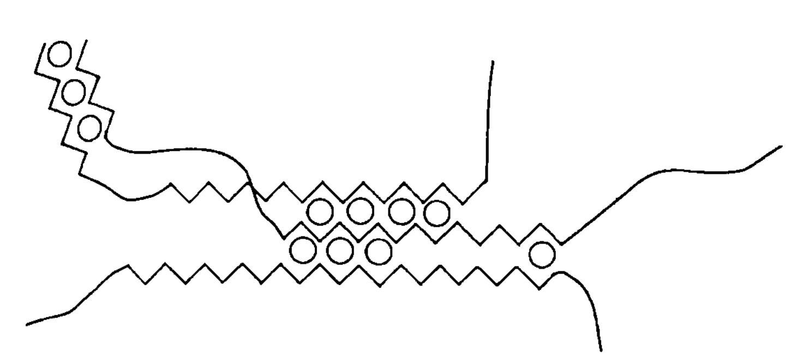

Figure 6 *Egg box-type bonding found, for example, between gluronic acid segments of alginate and calcium*
(Reprinted with permission from *Sugar Confectionary Manufacture*, E. B. Jackson, ©1999 Aspen Publishers Inc.)

aggregates with calcium (Figure 6). Obviously the gelling ability is proportional to the proportion of homogeneous guluronic segments present. The gels formed by calcium alginate are not thermoreversible. In cold water, alkaline alginates do not gel in the absence of calcium. Under these conditions they only act as thickeners by increasing the viscosity. In the presence of calcium or acid the product does build up a gel structure. Alginates might be used to thicken a fruit filling. One application that has been suggested is as a gloss and non-stick coating on liquorice products. The traditional gloss was mineral oil, which has been banned in foods. Similar glazing applications might be possible.

3.15.3 Carrageenan

This material is another seaweed polysaccharide. The name is believed to be of Irish origin. The ability of certain seaweeds to gel large quantities of milk was noted in several places, including the West of Ireland and Brittany. The Breton milk gel was called blancmange.

The commercial source of carrageenan is the following red seaweeds: *Euchema cottonii*, *E. spinosum*, *Chondrus crispus*, *Gigartina acicularis*, *G. stellata*, *G. pistallata*, *G. skottsbergii*, *G. chamissoi* and *Iradaea*. These seaweeds grow in Argentina, France, Morocco, Peru, Chile, The Philippines and Indonesia.

The carrageenan is first extracted by treating the washed seaweed with hot water. The seaweed is then crushed in the presence of alkali to extract the maximum amount of carrageenan. A diatomaceous earth is then used as a filter aid in filtering the hot extract under pressure. The product produced is a clear syrup. The carrageenan is then precipitated from this syrup with alcohol. The coagulated carrageenan forms into fibres. The fibres are then pressed and washed with strong alcohol to dehydrate them. The strong alcohol used is like the absolute alcohol found in laboratories – it has less water than the azeotropic mixture of alcohol and water. Consequently, the alcohol extracts the residual water from the carrageenan, thereby bringing the water content of the alcohol towards the azeotrope. The alcohol can then be recycled by vacuum distilling out the water. The alcohol used must be food grade. In the UK, duty would be chargeable on this material.

Carrageenan is chemically a sulfated polysaccharide consisting of galactose units. A common backbone exists in all the different fractions. The main chain is D-galactose residues linked alternately α-($1 \rightarrow 3$) and β-($1 \rightarrow 4$). The fractions are distinguished by the different number and position of the sulfate groups. A 3,6-anhydro-bridge can exist on the galactose linked through the 1 and 4 positions. The gelling carrageenans kappa and iota contain β-D-galactose 4-sulfate linked through the 1 and 4 positions. κ- and ι-Carrageenans differ in that latter contains an additional sulfate group on the 3,6-anhydrogalactose. These carrageenans are always found contaminated with the other. μ- and η-Carrageenan are the biological precursors of κ- and ι-carrageenan. The seaweeds have an enzyme that catalyses the transformation by eliminating the 6-sulfate group. Conveniently, the alkaline extraction used on carrageenan also expedites this transformation, thereby improving the quality of the product as a gelling agent.

Different species of seaweed yield different carrageenan fractions:

κ-carrageenan is found in *Euchema cottonii*
ι-carrageenan is found in *Euchema spinosum*
λ-carrageenan is found in *Gigartina acicularis*

κ-, ι- and λ-Carrageenan are found in *Chondrus crispus*, *Gigartina stellata* and *Iradaea*

κ-Carrageenan is a gelling agent that is soluble at 60–70°C. It reacts to produce a thermoreversible gel with milk protein. The gels tend to syneresis and tend to be breakable. A synergy exists between carrageenan and carob gum, *i.e.* the two together have more effect than both used singly. ι-Carrageenan dissolves at around 55°C and is a gelling agent. It reacts with milk protein to give elastic gels that do not undergo syneresis. Gels formed from ι-carrageenan are thermoreversible. In contrast, λ-carrageenan is cold soluble and is used as a thickener. It gives very weak gels with milk protein.

When carrageenans form gels the chains associate through double helices. Any μ- or η-carrageenan present inhibits this process.

3.15.4 Gelatine

Gelatine is produced by hydrolysing collagen, a connective protein of animals. The commercial sources of gelatine are normally cattle or pigs. The raw material is either bones or hides. Collagen can be hydrolysed under either acid or alkaline conditions. Different types of gelatine can not be mixed as they have a different isoelectric point. The different types of gelatine are referred to by their origin and the agent used for hydrolysis, *e.g.* acid pigskin gelatine. Limed ossein gelatine is normally produced from cattle bones.

In producing gelatine, skins are processed directly while bones are washed to remove the meat and fat residues. This produces dry degreased bone. The bones are then treated with hydrochloric acid. This removes the phosphates present, which are then precipitated with lime to produce dicalcium phosphate for use in cattle food. The remainder of the bones is the bone or ossein collagen. The collagen, whether bone or hide, is then hydrolysed with acid or alkali as required. In the alkaline hydrolysis process the collagen is steeped in a bath of lime for several weeks at ambient temperature. In comparison, acid hydrolysis only takes one day at ambient temperature. Next, the acid or alkali is washed away and any remainder is neutralised. The material is then cooked with hot water to liberate the gelatine. The solution contains about 6–7% gelatine and is referred to as an extraction. Repeated extractions are made until the raw material is exhausted. The highest quality gelatine is in the earliest extractions. If the starting material was pig skin the fat can be recovered at this stage. The extractions are then filtered followed by vacuum evaporation to 30–40% gelatine. The solution is then sterilised at 140°C, followed by crash cooling to obtain a jelly. The crash cooling minimises degradation of the gelatine. The concentrated gelatine jelly is then extruded using extruders similar to those used to make pasta. The extruded material is air dried to the final moisture content. The dried material is then ground as needed, followed by blending to the required specification. A typical product would be 14% moisture 84% protein and 2% ash. As gelatine is hygroscopic it needs to be stored so that it cannot pick up water. If the moisture content is allowed to rise to 16% mould growth can commence. Care needs to be taken when working with gelatine solutions as they form an excellent medium for bacterial growth. Gelatine is a product where the microbiological quality is very important. The performance of the gelatine would be impaired if it had suffered bacterial proteolysis; this is important as with an animal product there is always a risk of contamination with pathogenic bacteria.

The gelling quality of gelatine is obviously an important property to users. The gelling quality is normally measured using several non-SI empirical methods. Measures in use are grams, Bloom, Boucher units, FIRA degrees and jelly strength. Of these the most used is Bloom strength. It is possible to use less of a high bloom gelatine or more of a low Bloom gelatine to produce the same result. Table 11 shows the percentage of gelatine of other Bloom grades needed to give a similar strength to 100 Bloom gelatine. Table 12 gives the relation between Bloom strengths and concentration for equivalent jelly strength. Commonly, available Bloom strengths are 60 to 260, although higher strengths are available.

Gelatine prepared by acid hydrolysis is normally referred to as type A, while gelatine produced by alkaline hydrolysis is referred to as type B.

Table 11 *Relationship between a solution of 100 Bloom strength gelatin and equivalent jelly strength of other Bloom grades. Values are concentration (%) of gelatin required to give a similar jelly strength*

Bloom strength of gelatin (%)							
60	*80*	*100*	*140*	*160*	*200*	*225*	*260*
7.7	6.7	6.0	5.1	4.8	4.3	4.0	3.7
10.3	8.9	8.0	6.8	6.3	5.7	5.3	5.0
12.9	11.2	10.0	8.4	7.9	7.1	6.7	6.2
15.5	13.4	12.0	10.1	9.5	8.5	8.0	7.4
18.1	15.7	14.0	11.8	11.1	9.9	9.3	8.7

From R Lees and B Jackson, *Sugar Confectionery and Chocolate Manufacture*, Leonard Hill, Glasgow (1973)

Table 12 *Relationship between Bloom strength of various gelatins held in solution having equivalent jelly strength*

Bloom strength of gelatin (%)	Gelatin needed to produce equivalent jelly strength (%)				
	60 Bloom	*100 Bloom*	*160 Bloom*	*200 Bloom*	*260 Bloom*
60	10.0	12.9	14.9	18.3	20.7
80	8.6	11.2	14.2	15.8	18.0
100	7.8	10.0	12.6	14.1	16.1
120	7.1	9.1	11.5	12.9	14.7
140	6.6	8.4	10.7	12.0	13.6
160	6.1	7.9	10.0	11.2	12.8
180	–	7.5	9.4	10.5	12.0
200	–	7.1	8.9	10.0	11.4
220	–	6.8	8.5	9.5	10.9
240	–	6.5	8.1	9.1	10.4
260	–	6.2	7.8	8.7	10.0

From R Less and B Jackson, *Sugar Confectionery and Chocolate Manufacture*, Leonard Hill, Glasgow (1973)

The usual alkali employed is lime. The raw material for gelatine is tropocollagen, which is present in the original hides or bones. This protein consists of three polypeptide chains arranged in a triple helix. In contrast, gelatine consists of several free or interassociated chains, ranging in molecular weight from around ten thousand to several hundred thousand. On extraction, monomers (α-chains MW 100 000), dimers (β-chains) and trimers (λ-chains) and some lower order peptides are released.

A very important property of proteins is the isoelectric point. This is defined as the point at which the total negative and positive charges on the molecule are balanced. This point is where it is easiest to precipitate the protein. By analogy with the pH scale the isoelectric point is written pI.

Type A acid processed gelatine has a pI of 6.3–9.5; type B alkali processed gelatine has a pI of 4.5–5.2.

As gelatine is an animal product it is unacceptable to vegetarians. Some religious groups have problems with gelatine. Kosher gelatine is available. Gelatine made from fish has recently become commercially available.

Gelatine has the property that on standing it can set to acquire the structure of the original collagen. Gelatine gels form, provided the concentration is high enough and the temperature is low enough. Thus for any concentration of gelatine gel there will be a setting temperature. The thermoreversible nature of gelatine gels is useful in several ways. The product can give a melt in the mouth sensation, waste material can be recycled and it is possible to deposit the product hot and leave it to set on cooling. As well as its use as a gelling agent, gelatine can be used as a foaming agent. Proteins tend to stabilise foams. When a mixture containing gelatine is whipped it is possible to arrange that the mixture cools, thus setting the foam.

In use gelatine is pre-soaked, when it absorbs 5–10 times its own weight of water. The swollen gelatine will dissolve at 50–60°C. As gelatine can be hydrolysed by heating it above 80°C, trying to dissolve gelatine by boiling it or boiling a gelatine solution is just further hydrolysing the gelatine. As gelatine is not stable to acid, any addition of acid must be as late as possible in the process.

Gelatine produces some textures that cannot be produced otherwise. The photographic industry has been unable to find a substitute for gelatine in photographic film and paper. Gelatine also has some uses in pharmaceutical products, *e.g.* gelatine capsules. Table 13 gives some uses of gelatine in confectionery with the type and percentage of gelatine used. Gelatine does have a few minor uses in confectionery, such as sealing almonds in sugared almonds and as a granulation binder for pressed sugar tablets.

Gelatine can be used with other hydrocolloids such as pectin, agar, starch or gum acacia. Gelatine and gum acacia have been used in Rowntree's fruit pastilles for over a hundred years. Using a mixture of hydrocolloids can lead to difficulties. In the case of gum acacia and gelatine, if the conditions are wrong coacervates will form. Using a mixture of hydrocolloids allows a range of textures to be produced. As examples, gelatine and gum give a hard compact texture, a gelatine agar and pectin mixture will give a short brittle texture, while gelatine and starch give a texture between these extremes.

Gelatine is one of the most versatile of gelling agents. It is used as the gel in pork pies as well as in bakery fillings and icings. In addition to its use as a gelling agent it is used to make, and sometimes gel, foams and in some minor uses such as a sealing layer.

Table 13 *Some uses of gelatine in confectionary*

Use	*% Gelatine*	*Bloom strength*	*Function*
Jellies	6–9	175–250	gelling agent
Wine gums	4–8	100–150	gelling agent
Marshmallows	2–5	200–250	whipping agent
Fruit chews	0.5–2.5	100–150	whipping agent
Extruded aerated products	3–7	100–125	whipping agent

3.15.5 Gellan Gum (E418)

Gellan gum has only relatively recently been introduced as a gelling agent. At the time of writing it is not universally legal in foods. It is the extracellular polysaccharide produced in the aerobic fermentation of *Pseudomonas elodea*. The organism is fed a carbohydrate, *e.g.* glucose, with a nitrogen source and inorganic salts. The production system works under very carefully controlled conditions of aeration, pH and temperature. The broth produced is then treated with hot alkali. The gum is precipitated by treatment with hot propan-2-ol. The monomers in gellan gum are rhamnose, glucose and glucuronic acid in the ratio 1:2:1.

Gellan gum has been promoted as a suitable gelling agent for making fruit flavour jellies. It is particularly suited to this application as it is very stable even in acid conditions.

The gels are made by adding the gellan gum to water while shearing, followed by heating to 75°C, then adding ions, and cooling to set. The level of gellan gum can be as low as 0.05%. As gellan gum sets in the presence of ions, suitable salts must be present. Suitable ones that are permitted in foods are potassium, calcium, sodium or magnesium. Divalent cations, such as calcium and magnesium, will gel gellan gum at 1/25th of the amount of the monovalent cations sodium or potassium.

It is claimed that gellan gum is compatible with gelatine, xanthan gum, locust bean gum and starch.

At the time of writing gellan gum has been approved for food use in several countries. It will probably be approved everywhere in time.

Gellan gum had a slow initial take up by the food industry. Some bakery industry uses are in pie and bakery fillings, bakery icings, frostings and glazes. Gellan gum is often used with other gums and thickeners.

If the alkali step is removed a high acetyl gellan gum is produced. This type of gellan gum gives a more elastic thermoreversible gel similar to gelatine. This product might be more successful but it needs separate food approvals.

3.15.6 Gum Acacia also known as Gum Arabic E414

Gum acacia is the exudate of *Acacia senegal* trees. The trees grow on the edge of the desert, where they prevent the desert coming forward. The trees bind the sand together and retain moisture. Because the trees are part of the Leguminosae they fix some nitrogen into the soil. Properly managed the trees provide a cash crop, a source of firewood and maintain the fertility of the soil. Some gum is cultivated but a little comes from wild trees. Traditionally, the best gum comes from the area around Kordofan in the Sudan. Most gum acacia comes from the Sudan, although other countries such as Chad, Senegal and Niger produce acacia gums.

Gum acacia is a unique polysaccharide, with some peptides as part of the structure and has a range of different uses. It was originally the gum in gum sweets although some gum sweets do contain modified starch as a substitute. The replacement of gum is not because the substitute performs better but because there have been supply problems with gum acacia. Gum acacia is likely to be encountered in bakeries in small quantities when it has been used to make emulsions of citrus oils as a bakery flavour. It is possible to use gum acacia in making dry flavours from oils such as citrus by making an emulsion and then spray drying it.

Traditionally, the best gum was produced by hand picking clear tears of gum. This grade is still available but the price reflects the cost of hand picking. As the gum is produced by removing the tears from the gum trees some gum is contaminated with pieces of bark. The gum can pick up colour and astringent tastes from the bark. In practice raw gum is often contaminated with desert sand. In confectionery use the light coloured grades are used to make products that need a light colour while darker gum is used in products that have dark colours and flavours, *e.g.* liquorice.

There is a need to have a testing regime to ensure that the gum acacia offered is gum acacia and not a product from some other species that is unsuitable. Acacia Seyal gum is sometimes encountered, which is less soluble than gum acacia and hence it is unsuitable for making sweets with a high proportion of gum acacia as it will not dissolve sufficiently. Instances have occurred where gum combretum, a product that is not an acacia gum, has been found in commercial supplies purporting to be gum acacia.

Raw gum acacia tends to arrive with natural contamination of bark and sand. The raw gum is normally purified by first removing any stones and then dissolving the gum. The insoluble contaminants are removed by filtering or centrifuging the insoluble material. A typical process would involve adding the gum to water and heating gently to produce a solution of from 30 to 50% gum. Gum acacia is much more soluble than other gums. It is possible to make a 50% solution in cold water. The viscosity of a gum solution falls with increasing temperature as well as being pH dependent. Maximum viscosity occurs at pH 6 but falls above pH 9 and below pH 4.

Some gum users now take gum in a pre-prepared form. Spray dried gum acacia has been used in pharmaceutical products for some time. The spray dried gum offers the pharmaceutical manufacturer a clean ready to use product. Instant forms of gum acacia have been offered by suppliers for some time. The instant products can be rapidly made into solution and used. Obviously the instant gum is more expensive. A manufacturer that uses gum as a minor ingredient may well find that the capital and labour cost of purifying raw gum is not cost effective. A company that uses gum acacia as a major ingredient might come to a different conclusion. Instantised gums pose different problems to the analytical chemist. One approach that can be used is to have an optical rotation specification for the product. Even this approach is not entirely proof against a material that contains a blend of gums of different optical rotation.

3.15.7 Guar Gum

This vegetable gum comes from *Cyamopsis tetragonolopus*; it is extracted from the endosperm of the seeds. The countries of origin are India and Pakistan.

An impure product, guar flour, is made by milling the endosperms, having first de-hulled them. The resulting product is obviously impure and gives a cloudy aqueous solution.

The pure product is produced by dissolving the gum from the seeds in hot water. Diatomaceous earth filtration is then used to purify the solution. As the gum is less soluble in alcoholic than aqueous solutions it is precipitated by adding propan-2-ol. The pressed filter cake is then washed in pure alcohol to dehydrate it. The alcohol is then recovered by pressing again. The pressed product is then milled to the required final size.

Chemically, guar gum is a galactomannan, *i.e.* it is composed of β-D-mannose and α-D-galactose units. The molecule has a main chain composed of $(1 \rightarrow 4)$ linked β-D-mannose residues with side-chains of $(1 \rightarrow 6)$ linked α-D-galactose. Guar gum is chemically very similar to locust bean gum and was originally developed to make up for a shortage of locust bean gum. The differences between the two gums are in the number of galactose molecules attached to the mannose chain. Guar gum has a ratio of D-galactose to D-mannose of 1:2 while the same ratio in locust bean gum is 1:4. While locust bean gum does not itself gel it does form gels with carrageenan. Locust bean gum is a possible minor ingredient in bakery products.

3.15.8 Pectin

Pectin is used in foods in two forms, high methoxyl pectin and low methoxyl pectin. High methoxyl pectin is the form normally found in fruit while low methoxyl pectin is a chemically modified pectin. Pectins are acidic polysaccharides that occur in the cell walls of fruit. The commercial source of pectin is either citrus peel or apple pomace. The citrus peel is the residue from the production of citrus juices while apple pomace is the residue of cider production. Thus pectin is a by-product of either cider or fruit juice production.

Pectins are extracted from the raw material by using hot hydrochloric acid. The acid treatment hydrolyses the pectin from the protopectin. Pressing and filtration with a filter aid are then used to remove the insoluble material. The next stage in the process is to precipitate the pectin. The way in which the pectin is precipitated depends on the type of pectin that is being manufactured. A rapid set high methoxyl pectin would be precipitated as soon as possible. A lower degree of methoxylation is obtained by holding the extract for some days, which removes some of the methoxyl groups. If amidated low methoxyl pectins are being made the pectin is treated with ammonia at this stage. In the next stage the pectin is precipitated with alcohol. The resulting precipitate is washed successively with alcohol of higher strength, finishing with pure alcohol. The pure alcohol removes the residual water as it has a higher affinity for water than the pectin. This gives a fibrous pectin that is then dried, ground and sieved. An alternative method of precipitating pectin is to treat it with aluminium ions to produce an insoluble aluminium pectin salt. The aluminium is subsequently removed by treating with acidified alcohol.

Chemically pectins can be regarded as a polygalacturonic acid. Typical molecular weights are between 2000 and 100 000. The pectin molecule is a polymer with galacturonic acid monomers linked through $1 \rightarrow 4$ bonds. Some of the galacturonic acid groups will be methoxylated. The ratio of methoxylated to unmethoxylated galacturonic acids is referred to as the degree of methylation, normally abbreviated to DM. This is an important parameter as it controls how the pectin behaves as well as how it is treated in food legislation. The DM is defined as the average number of methoxyl units per 100 galacturonic units. Pectins with a DM above 50 are referred to as high methoxyl while those with a DM below 50 are classed as low methoxyl pectins. The low methoxyl pectins are sometimes further modified by converting some of the acid groups into amides by treating the pectin with ammonia. The degree of amidation is the average number of amide groups per hundred galacturonic units.

Most countries restrict the maximum degree of amidation to a 25% maximum. High methoxyl pectins are naturally present in fruit and escape restrictions on use for that reason. Low methoxyl pectins are treated as additives and have restrictive acceptable daily intakes (ADIs).

High methoxyl pectins are used for making low pH products such as fruit jellies. The setting conditions for high methoxyl pectins are a high soluble solids and a low pH. The high methoxyl pectins are further subdivided by their speed of gelation. The speed of gelation is controlled by the DM. A DM between 68 and 72% gives a rapid set while a DM between 66 and 70% gives a medium set. A slow set is achieved with a DM between 59 and 64%. In confectionery products the soluble solids are so high that only slow set pectins are used. A rapid set pectin would pre-gel under these conditions.

Low methoxyl pectins have radically different properties, *i.e.* a small chemical modification has totally altered the way in which pectins behave. The low methoxyl pectins are set by calcium ions independently of the pH. Because hard water normally contains calcium ions care must be taken in selecting low methoxyl pectins when using hard water supplies or when moving recipes between factories. The gel produced in low methoxyl pectins has the egg box structure found in alginates (see Figure 6).

The gel properties as well as the gelling conditions are radically different for the two types of pectin. High methoxyl pectins produce a gel that does not remelt, while some low methoxyl pectin gels are thermoreversible.

The best established use of pectin is in making jam. While some fruit have sufficient pectin and acidity to make a well set jam others, *e.g.* strawberries, benefit from the addition of pectin from another fruit.

Another use of pectin is when making a fruit flan or an open tart. These products are often coated with a pectin jelly based on a high methoxyl pectin. The pectin is dispersed and heated to dissolve it. As high methoxyl pectin requires an acid pH to set, just before use acid is added and the pectin mixture is poured over the flan. These fruit products are expected to be acid as part of the fruit flavour, so an acid gel is acceptable.

If a product with a neutral pH is being made a high methoxyl pectin would not set, therefore the only option is to use a low methoxyl pectin. Examples of products with a neutral pH are mint flavoured jellies and Turkish delight.

Pectin does have some compatibility with other gelling agents. In particular it is used in conjunction with gelatine. Pectin suppliers claim that up to 25% of the gelatine can be replaced without significantly altering the texture. The benefit of the replacement is of course purely financial. Higher levels of replacement will cause the texture to become softer and less chewy. Pectin is compatible with starch but gives a pasty texture that is not very popular. This texture does seem to be acceptable in Turkish delight. The problem with mixing hydrocolloids is that if the pH is wrong the entire system will become unstable. The stability of the hydrocolloid in solution depends on the electrostatic charge on the molecule. Anything that neutralises that charge is likely to bring the material out of solution.

In working with high methoxyl pectin the pH must be controlled because below pH 4.5 slow-set pectins degrade, causing a loss in gel strength. At pHs below 3.2 there is a danger that the pectin would pre-gel when the solution reached final strength. It is normal to use a buffer such as citric acid and potassium citrate to prevent these problems. Pectins premixed with suitable buffer salts are commercially available. This type of product avoids having staff skilled enough to accurately make up buffers. This is attractive to small manufacturers.

One problem with high methoxyl pectin is that of using rework. As pectin gels do not remelt this is much harder than it would be with gelatine. This point is important because not only is some of an expensive ingredient lost but simply disposing of the waste produces another cost. In practice high methoxyl pectin rework can be used if it is first comminuted, but the proportion used must not exceed 5%.

Pectin has been suggested as an ingredient for aerated products. A compatible whipping agent would be used in conjunction with a high methoxyl pectin. Typically, the product would contain 0.5–2.5% of high methoxyl pectin and some gelatine. Low methoxyl pectin is well suited to making jellies with a neutral flavour. The commonest examples are vanilla, peppermint or rose water (used in Turkish delight). These products are likely to have a pH in the region of 5. This sort of pectin produces a softer and less elastic texture than the high methoxyl pectins. The gelling conditions for low methoxyl pectin are completely different to those for high methoxyl pectin, being from pH 2.8 to 6.5 and from 10 to 80% soluble solids in the presence of calcium. The ability to trigger the gelling mechanism by adding calcium has led to several innovative ideas. One is to use the gel to make crust-less liqueurs; another is to make liquid-centred fruit flavoured products. In both cases the product relies on the pectin gelling on contact with calcium ions.

3.15.9 Starch

Starch is the major energy storage polysaccharide of cereal crops. It is a natural polymer of dextrose. Starch has two naturally occurring forms: one is amylose, a polymer with long linear chains, and the other is amylopectin, a branched-chain polymer. The length of the amylose chain varies between different plants but the common values are between 500 and 200 glucose units. Although the chain lengths in amylopectin are limited to 20–30 glucose units it tends to be a more massive molecule than amylose. Different plants have different ratios of amylose to amylopectin – indeed this is responsible for most of the variations in starch properties between the starch from different plants. Virtually any starch-containing crop could be used as a source of starch. Sources that are used commercially are maize (US corn), wheat, potato, rice tapioca, or sago. The common types of starch are extracted from maize, wheat or potatoes. The other crops are used when starches with special properties are required. The choice of raw material ultimately depends on economics and availability. Wheat starch can be produced as a by-product of the production of dried wheat gluten.

The methods used to separate the starch vary, depending on the raw material. Maize is normally wet milled. Initially the maize kernels are steeped in dilute sulfuric acid for 40–50 hours to soften the kernels. Next, the kernels are milled to release the germ that contains the oil. The fibre is then separated from the endosperm by milling it finer. Centrifuges are then used to separate the starch from the protein. After this the starch is washed and dried.

The variation between the starch from different plants is considerable. The percentage of amylose varies from 27% in maize starch through 22% in potato starch to 17% in tapioca starch. The waxy maizes are unusual in that they are almost pure amylopectin. This is extremely convenient because it avoids the need to separate amylopectin from amylose chemically.

3.15.9.1 Cooking Starch. A fundamental difference between starch and the other gelling agents is that starch has to be cooked rather than dissolved. Indeed, raw starch is insoluble. When starch is examined under the microscope it can be seen to consist of discrete granules. The shape of the granule depends on the origin of the starch.

These granules contain micelles of starch molecules. When the granule is heated in water at a given temperature the granules swell and start to absorb water. This process is called gelatinisation and the temperature is known as the gelatinisation temperature. This temperature is a characteristic of the different types of starch. Maize starch gelatinises from 64–72°C. There is an exception to this and that is waxy maize starch, which forms non-gelling clear fluid pastes. Waxy maize starch behaves as a gum rather than a gelling agent. This is one of the types of starch that are used as gum acacia substitutes.

The changes that occur in a starch when it is heated in water can be studied in several ways. One way is to follow the changes under a microscope, another is to measure the viscosity of the paste. A Brabender amylograph is normally used to monitor the paste viscosity.

On cooking maize starch the viscosity increases when the starch begins to gelatinise. As the temperature rises towards 95°C the viscosity falls. When the paste is cooled the viscosity rapidly increases. The variation of viscosity with temperature is characteristic for each different origin of starch. Potato starch, for example, has a lower gelatinisation temperature than maize starch but has a higher maximum viscosity. When cooled the viscosity of potato starch rises less. Once again amylopectin starches do not show this behaviour as they do not gel.

3.15.9.2 Obtaining Different Properties in the Starch. One method of obtaining a starch with different properties is the biological method of using a different type of plant. The best example of this is waxy maize, which yields a starch that is nearly pure amylopectin. The other method is to chemically modify the starch. Chemically modified starch is normally declared as "modified starch". A whole range of modified starches are available. There is of course no bar to chemically modifying a starch from a special source.

3.15.9.3 Pregelatinised Starches. These starches have been gelatinised either by extrusion or by heating in water followed by roller drying.

3.15.9.4 Oxidised Starches. The effect of oxidation is to diminish the tendency to form micelles, which in turn reduces the tendency to gel, as well as making the paste more stable. The usual oxidising agent is hypochlorite.

3.15.9.5 Non-gelling Starches. These products are intended for uses where the starch replaces a gum like gum acacia. A typical product for this use might be an oxidised waxy maize starch.

3.15.10 Locust Bean or Carob Bean Gum

This material is another plant polysaccharide. The source is the seeds of the carob tree (*Ceratonia siliqua*), also known as the locust bean tree. The trees grow around the Mediterranean and in California. An alternative name for the fruit is "Saint John's Bread". An impure material called carob pod flour can be produced by just removing the hulls and milling the endosperms directly. An impure product like this will give a cloudy solution in water. To produce a pure material the gum is dissolved from the seeds with hot water. The gum solution is then purified by diatomaceous earth filtration. Next the gum is precipitated by adding propan-2-ol. The resulting precipitate is pressed, followed by washing with pure alcohol to dehydrate it. Following this the product is pressed to recover the alcohol and then dried and milled to the required size.

Chemically the gum is a galactomannan composed of β-D-mannose and α-D-galactose units. The β-D-mannose units are (1→4) linked to make the main chain while the α-D-galactose units form the side-chain. Locust bean gum is very similar to guar gum. Indeed, it was developed as a substitute when locust bean gum was not available. The two gums differ in the ratio of D-galactose to D-mannose. In guar gum the ratio is 1:2 while in locust bean gum the ratio is 1:4, *i.e.* the difference lies in the number of galactose residues attached to the main D-mannose chain.

Locust bean gum on its own is a thickener. It will dissolve in water at 80°C. When used with κ-carrageenan the substances exhibit synergy in producing an elastic and very cohesive gel. A similar synergy occurs with xanthan gum, again producing an elastic and very cohesive gel. Locust bean gum is also used to stiffen agar jellies. In general, this gum is too viscous on its own to be much used in confectionery.

3.15.11 Xanthan Gum

This gum was the first microbial gum to be used in the food industry. It is produced by the aerobic fermentation of *Xanthomonas campestris*. A specially selected culture is grown on a carbohydrate-containing nutrient medium with a nitrogen source and other essential elements. The pH, temperature and aeration are controlled carefully. The product is then sterilised and the gum is precipitated with propan-2-ol. Next, the precipitate is washed, then pressed to remove residual alcohol, followed by drying and grinding to the required size.

Chemically xanthan gum is an anionic polysaccharide with monomers of D-glucose, D-mannose and D-glucuronic acid. The polymer backbone is composed of (1→4) linked β-D-glucose units, similar to cellulose. On alternate glucose units a trisaccharide chain containing one glucuronic acid and two mannose residues is fixed to the 3-position. This gives a stiff chain that can form single, double or triple helices. The molecular weight is approximately 2×10^6. The molecular weight distribution is narrower than for most polysaccharides. Xanthan gum solutions behave as though they are a complex network of entangled rod-like molecules.

Xanthan gum dissolves in cold water. A 1% solution has a pH between 6.1 and 8.1. It normally functions as a thickener but combines synergistically with locust bean gum to produce a very cohesive and elastic gel. Xanthan gum is used in making gluten-free bread but it is one of the few substances that can be used as a substitute for gum tragacanth.

3.15.12 Egg Albumen

In this work the name egg albumen is used to refer to the mixture of proteins in egg white. Around 54% of the protein in egg albumen is ovalbumin. The number of proteins identified in egg albumen will continue to increase with improved analytical techniques. Unless fractionated egg albumen proteins become available this will be of little consequence for confectionery makers.

3.15.12.1 Practical Forms of Egg Albumen. Fresh egg white is not normally used in bakery products. The fresh product is too unstable and could have bacteriological problems. In practice, various forms of dried albumen are used. These have the advantage that they can be thoroughly tested bacteriologically before use. The form most commonly used in confectionery is dried egg white. Typically this is made by pouring egg white into shallow trays and drying it. The resulting sheets are then ground to final size. This is a low technology product. It is possible to apply sophisticated drying methods, *e.g.* spray drying to produce egg albumen that will reconstitute to a product similar to fresh egg white. "Fluff dried" albumen is made by whipping the albumen followed by rapidly drying the resulting foam. The sophisticated forms of egg albumen do give superior results in some bakery applications. The anecdote is often repeated, particularly by marketing men, of the cake mix that sold better when the instructions were changed to "add an egg". In practice a cake mix formulated from old fashioned dried egg albumen will produce a rather unsatisfactory cake. In this application one of the more sophisticated egg albumen products is needed and is worth the extra cost.

3.15.12.2 Properties of Egg Albumen. Egg albumen is normally used in foods for two reasons: it whips into a foam and the foam can be set irreversibly by heat. One advantage of egg albumen is that it is relatively unaffected by the presence of fat. Fat in these systems acts as a foam breaker. Some other whipping agents are badly affected by the presence of fat. In confectionery systems egg albumen is usually set by beating the reconstituted egg albumen into a hot sugar syrup. The coagulation temperature is affected by the water activity. For example, in a 40% sugar syrup the coagulation temperature is 65°C, but in a 60% sugar syrup the coagulation temperature rises to 75°C. One advantage of egg albumen is that it always coagulates reliably.

3.15.12.3 Testing Egg Albumen. Egg albumen needs to be tested thoroughly. It needs to be tested chemically to see that it is not contaminated with heavy metals, as do all food materials. It needs to be rigorously tested for microbiological contamination. This is particularly important as there is a risk of salmonella contamination. Egg albumen also, unlike some other materials, needs to be tested to see that it will actually perform satisfactorily in the product in question. A sample of egg albumen that would be entirely satisfactory for making nougat might be unsatisfactory in a cake mix. This could obviously be done by making an experimental batch of product. Often this is not convenient, particularly where making a small batch is not easy or where the product time is long. This problem has been tackled by devising various empirical tests. In theory, an empirical test should work even if the theory of its operation is not understood. However, empirical tests have, upon examination, been found to be of low predictive value. A typical empirical test might be to mix a given weight of egg albumen in a given volume of water on a particular mixer for a specified time. The resulting foam would be poured into a funnel with a weighed tube under the spout and the height of the foam measured initially and after a length of time. The weight of liquid that runs into the tube is then measured. The results of this sort of test are then compared with the specification. The egg albumen would then be rejected or accepted accordingly. In one case the test protocol specified a particular model of food mixer. When this mixer was discontinued the testers were forced to look around for an alternative type of mixer and to recalibrate the test.

The variation in the performance of different types of egg albumen in different systems is almost certainly caused by variations in the degree of denaturation of the protein. Those products that work best with fresh egg white clearly need an undenatured product.

3.15.12.4 Substitutes for Egg Albumen

Milk Proteins. As some milk proteins will gel on heating and others can be modified to make whipping agents it has long been thought that milk proteins could be used as whole or partial substitute for egg proteins. Purified whey proteins were regarded as a suitable raw material as whey is a low value by-product from cheese making. Early products in this area were not very successful. Residual fat in the product inhibited foaming. A more serious problem was that active lipase enzymes in the ingredient were introduced into the finished product. Lipases break fats down to their constituent fatty acids. In the case of butter fat the principle product would be butyric acid. At low levels butyric acid gives a pleasant buttery or creamy flavour. At higher levels the flavour changes to cheese, ultimately ending up as Parmesan cheese. Cheesy flavours are not popular in sweet products. If the product is made with a lauric fat or contains nuts that usually contain lauric fat the lipase will liberate lauric acid. As soap normally contains sodium laurate it is not too surprising that free lauric acid tastes of soap.

Satisfactory products based on milk protein have been produced. One such is the range of products sold under the trade mark Hyfoama. This material is normally used in conjunction with other gelling or whipping agents. Typically, Hyfoama is used with egg albumen. The properties of the two substances complement each other. Hyfoama foams reliably but does not coagulate reliably. In contrast, egg albumen always coagulates reliably but sometimes foams badly. All foaming agents are sensitive to the presence of fat and Hyfoama is no exception.

Soya Proteins. Early attempts to make albumen substitutes from soya protein also ran into problems. A bean flavour tended to appear in the finished product. A solution to these problems has been found. Whipping agents based on enzyme modified soy proteins are now available. The advantage of enzymatic modification is that by appropriate choice of enzymes the protein can be modified in a very controlled way. Chemical treatment would be far less specific. In making these materials the manufacturer has control of the substrate and the enzyme, allowing the final product to be almost made to order. The substrates used are oil-free soy flakes or flour or soy protein concentrate or isolate. The enzymes to use are chosen from a combination of pepsin, papain, ficin, trypsin or bacterial proteases. The substrate will be treated with one or more enzymes under carefully controlled conditions. The finished product is then spray dried.

In use these soya based products, unlike egg albumen, do not coagulate. They must be used in conjunction with egg albumen or another coagulating material if coagulation is needed. The soya-based proteins have the advantage that they have approximately twice the whipping capacity of egg albumen. The modified soya protein is used by dispersing it directly in two to three times its own weight of water. It is not necessary to pre-soak this material, unlike egg albumen.

CHAPTER 4

Analytical Chemistry

4.1 INTRODUCTION

Notably, the two most important points in any analysis are the provenance of the sample and whether the sample is representative of the bulk. If either of these points is not attended to, however sophisticated or diligent the analysis, the results are at best pointless or even positively misleading.

Apart from non-viscous liquids the importance of obtaining a representative sample can not be overstressed. As an example flour contains very wide range of particle sizes and several different substances. It is possible to air classify flour to extract the protein. If a sample is not representative the values obtained will not be reliable.

4.2 METHODS

The analytical methods given in the following sub-sections are likely to be of interest to those involved with the bakery industry. Chemists once classified methods of analysis into physical methods and chemical (wet) methods. There was even a feeling that physical methods were in some way inferior.

The time pressures of modern industry are such that rapid methods of analysis are a necessity. Physical methods of analysis have a big advantage in the food industry as chemical methods must be carried out in a laboratory separated from the production facility.

The ultimate improvement from a production control view is to perform the analysis on-line. This speeds things up further and allows almost instantaneous testing of the material.

4.2.1 The Kjeldahl Method

The Kjeldahl titration remains the chemical method for determining the nitrogen and hence protein content of wheat or flour. The method works by assuming that any nitrogen present in the sample is present as protein. Once the nitrogen has been measured the quantity of protein present is calculated by multiplying the nitrogen content by an appropriate factor for the type of protein.

A weighed sample is boiled in concentrated sulfuric acid, which quantitatively produces ammonia that reacts with the excess sulfuric acid to produce ammonium sulfate. An excess of sodium hydroxide is then added and the liberated ammonia is distilled into an excess of a standard acid solution, which is then titrated with sodium hydroxide.

Rather than perform this analysis in laboratory glassware in a fume cupboard, special pieces of apparatus that hold the flasks and allow several Kjeldahl titrations to be carried out in parallel are employed.

The Kjeldahl method is not a rapid means of analysis but it does have the advantage of being absolute. It is a sobering thought that a batch of bread can be made in less time than it takes to check the protein content of the flour by Kjeldahl.

4.2.2 Near-infrared Spectroscopy

Spectroscopists divide up the electromagnetic spectrum on the basis of the techniques used to obtain measurements (Figure 1). The near-infrared (NIR) region is the part of the infrared that is nearest to visible light. In the near-infrared region the technology used is very similar to that used in the visible, *e.g.* filament lamps can be used as light sources. The next region to the near-infrared requires more specialised instrumentation and is the region used by organic chemists to identify the presence of chemical groups in compounds, it is normally called the mid-infrared. The far-infrared requires very specialised techniques.

The atoms in molecules can be regarded as behaving like two spheres connected by a spring, with a natural vibration frequency. It so happens that the vibrations of atoms tend to correspond to an energy equivalent to a wavelength in the mid-infrared. It would appear that mid-infrared wavelengths would be useful for measuring the quantity of substances in systems like flour. In practice that is not so. Infrared peaks have a lot of

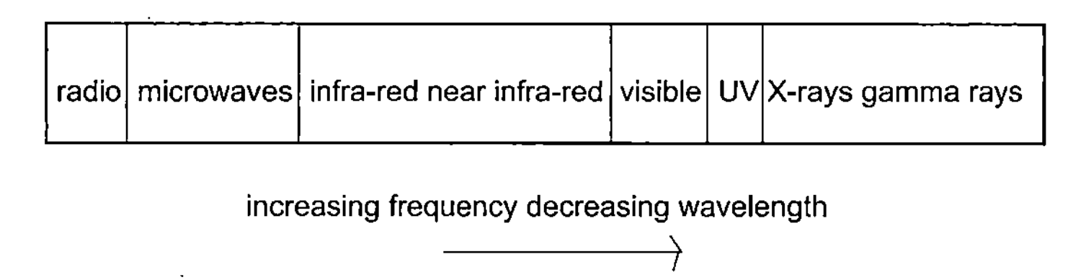

Figure 1 *The electromagnetic spectrum (note the position of the near-infrared)*

fine structure and for several spectroscopic reasons it is not normally possible to obtain quantitative results.

However, any vibrating system not only has a natural vibration frequency but will also vibrate at twice that frequency, which is known as the first overtone. The first overtone of the vibrations of molecules like water, proteins and fats correspond to a frequency in the near-infrared. Because these frequencies are overtones all of the spectroscopic problems that preclude making quantitative measurements in the mid-infrared are not present in the near-infrared.

The problem with near-infrared spectra is in assigning a wavelength to a particular chemical group. In general, spectra in the mid-infrared give sharp peaks that have much fine structure but can easily be assigned to particular chemical groups. Near-infrared spectra, however, resemble a range of distant hills with some hills a little taller than the others. Apart from water, it is not immediately apparent which is the best wavelength to use for the constituents. Initially, wavelengths were chosen on the basis of scanning samples and attempting to find correlations between the reciprocal of the reflectance and the composition of the samples. Some of these early calibrations definitely used the wrong wavelengths. As there is normally a correlation between the water content of a sample of flour and its protein content some early protein calibrations were in fact using wavelengths associated with water. Most of the time this would work but a high protein low moisture sample would not give reliable results. The usual correlation between the protein content and the water content occurs because protein absorbs water.

Current calibrations for water, protein and fat use wavelengths that are known to correspond to the vibrations of –OH, –NH and –CH groups.

Thus the absorption of infrared energy by solids is an approximation of the Beer–Lambert exponential law for transmission:

$$I_t = I_0 \exp(-kx)$$

where x is the path length for the radiation, k is an absorption coefficient, I_0 is the infrared energy incident on the sample and I_t corresponds to the transmitted energy.

This can be rearranged as:

$$\log(I_t/I_0) = kx$$

This gives a simple relationship between the amount of the absorber and the ratio of the incident and the transmitted energies. Now these equations apply to transmission spectroscopy, which is sometimes used for on-line measurements, but the commonest NIR measurements are made by reflectance. In reflectance measurements the additional problem is the effect of light scattering.

Modern instruments give very reliable results.

4.2.3 Fat Content

The fat content of products and ingredients can be measured by either wet chemical methods or instrumental methods. The normal wet chemical method is the use of a Soxhlet extractor and petrol.

4.2.4 Chromatography

In 1904 the Russian scientist Tswett coined the term chromatography (= colour writing) with regard to his work in separating plant pigments. Today, some of the most potent analytical methods available are chromatographic.

Chromatography using glass columns with a gravity flow is still carried out by biochemists using soft carbohydrate packings.

4.2.4.1 GLC or GC. The acronyms stand for gas–liquid and gas chromatography. The original packings were either a solid or a solid coated with a liquid. The mobile phase is a gas, hence the term gas or gas–liquid chromatography. The liquid is usually forgotten.

Gas chromatography is one of the most powerful analytical techniques available. Its only major limitation is that it can not analyse involatile compounds such as fats. The solution in this case is to make a volatile derivative, *e.g.* the use of fatty acid methyl esters to analyse triglycerides.

4.2.4.2 HPLC. This acronym is variously expanded as either high-pressure or high-performance liquid chromatography. The cynical refer to it as high-price liquid chromatography.

As chromatography takes place at a surface, performance can be increased by using smaller particles, which have more surface. Smaller particles lead to higher back pressures, hence the high pressure.

HPLC has many applications in the food industry. Two of interest in the bakery sector are the analysis of triglycerides and wheat proteins.

4.2.4.3 Ion Chromatography. Ion chromatography is a specialised technique that uses high-performance ion exchange columns. Its major use in the food industry is the analysis of sugars.

CHAPTER 5

Flour Testing

5.1 INTRODUCTION

Typically, flour is tested in bakeries to determine if it can be used to make a particular product, to find if it is fit to use and to determine if it is the material specified. Simply supplying the wrong product by accident can have devastating consequences on a bakery.

Tests applied to flour can be divided as follows.

5.1.1 Analytical Tests

For example, measuring the protein content.

5.1.2 Empirical Tests

These tests measure the way in which flour behaves when it has been made into a dough. They work with a specially made dough and give an indication of how the flour will behave in that situation. This sort of information is inherently more useful than a mere protein content.

5.1.3 Test Baking

In theory, analytical testing ought to be able to answer all pertinent questions but, unfortunately, it can not. While measuring the protein content will discriminate between a low protein flour and a high protein flour, the protein content will not necessarily guarantee that a given flour will make a satisfactory loaf of bread. The problem is that it is much easier to measure the quantity of protein present rather than its quality.

Test baking is one answer to this problem; some flour samples that had a good measured composition produce a poor loaf of bread. There are also samples that do not have very promising measured properties but bake well. The problem usually can be explained in terms of protein quality. Flour that has a high protein content and a high falling number normally bakes well. An exception would be flour made from English wheat where the wheat has been overheated in drying it. This damages the proteins and leads to poor performance in the bakery.

The situation is worse where the final product is not bread but some other product, *e.g.* biscuits. Analytical measurements performed on flour are aimed at measuring its suitability for making bread. For biscuits, the protein content is not important except that for most biscuits high protein flour will not give a satisfactory product.

In general the properties of a good bread flour and a good biscuit floor are almost opposite. A good bread flour will have a high protein content, a high Hagberg Falling Number and high starch damage. A good biscuit flour will be low in protein with a low starch damage. A high Hagberg Falling Number is of no advantage.

Baking tests have to be the ultimate determinant of a flour's suitability for a given purpose. If the flour can be made into a satisfactory product in a test bake then it will work in the bakery. If the test bake is unsatisfactory then no brandishing of analytical results will make the flour fit for the use.

While test baking can provide information that laboratory measurements can not it is time consuming and not always convenient. Other methods of flour testing have been developed to bridge the gap between the analytical results and the test bake. Some of these methods use specialised equipment that makes and performs carefully controlled tests on a dough, *e.g.* the Brabender Extensograph. The choice of testing regime varies from country to country and in some cases from company to company. As an example, in French speaking markets the Chopin Alveograph is the testing regime of choice.

Some empirical testing regimes, particularly those used to test the suitability of flour for products other than bread, have come about because of a need for a rapid test to show that a flour would work in a specific plant. Some cases require a test that can distinguish between flours that have a similar analysis but where only some of the flours would be satisfactory.

One such test involved the use of a domestic table mixer to detect flours that would suffer gluten separation when used on the plant. The test involved making a standard batter in the mixer and then beating it at maximum speed to see if gluten separation occurred. A table mixer is obviously much cheaper than a precision scientific instrument. Unfortunately for the company concerned, the mixer manufacture modified the speed control system of later versions of the mixer, such that results on different models of the same mixer were not directly comparable. The mixer manufacturer of course was making a mixer and not a precision instrument!

Empirical tests are chosen by end-users because they have been shown to produce useful information. Some empirical tests have been elevated to the status of standard tests. Others remain in use by their inventors.

5.2 EMPIRICAL TESTING REGIMES

5.2.1 The Hagberg Falling Number

The Hagberg Falling Number test only just qualifies for this section as it is an effective way of measuring the α-amylase activity of wheat or flour. It has the considerable advantage that it is unaffected by any added fungal α-amylase. In addition, it only requires a supply of distilled water and electricity – there is no need for chemical glassware or any reagents.

The major problem when carrying out the test on wheat is not one of methodology but concerns the source of α-amylase. α-Amylase levels are highest in sprouted grains of wheat. If a sample contains one sprouted grain of wheat the falling number will be sufficiently low that the wheat will be failed for bread making. Another sample from the same batch without a sprouted grain might pass.

When wheat is being tested it must first be ground, but flour is tested directly. In use a weighed quantity of the flour is shaken in a precision-made test tube with a volume of water. A stirrer of standard design is used. The tube is then placed in the apparatus where the stirrer is held up. The apparatus contains a boiling water bath, which gelatinises the starch in the flour. After 60 seconds the stirrer is released and the time it takes to fall is recorded. The falling number is this time plus 60 seconds. Flour for bread making should ideally be above 250 falling number or at least 220. In a bad year bread can be made with flour at 200 falling number. High falling numbers are not necessarily an advantage in all products. In biscuit flours and flours for some pizzas a lower falling number is more desirable.

When Canadian wheat flour was the norm in British bakeries, with a falling number as high as 600, it was desirable to introduce malt flour to increase the α-amylase to feed the yeast and open up the structure of the crumb.

In a year when there has been rain during harvest, causing wheat to sprout, the falling number becomes very important. An example of such a year was 1987, when rain in East Anglia meant that the winter wheat in that area was largely unfit for bread making. In contrast the spring wheat from East Anglia was of good quality that year. Unfortunately, not much spring wheat was sown and the deficiency had to be met from other areas of the UK and Europe.

The Hagberg Falling Number is a test that is used across Northern Europe, where excessive amylase values can be a problem. However, it is not used in Spain, presumably because the climate is such that excessive amylase does not occur.

5.2.2 Chopin Alveograph

This instrument is used in France and French speaking countries to test the baking quality of wheat and flour. A flour water dough is made under standardised conditions and rolled into a sheet. Four discs are then cut from the dough and are allowed to rest for 20 min in a constant temperature chamber. One of the discs is then clamped across the orifice of the Alveometer and air is passed through the orifice, blowing a bubble in the dough. This continues until the bubble bursts. The Alveograph records the pressure variation with time. Three parameters are derived from these curves: The maximum pressure, the time the bubble took to burst and the area under the curve. The results of the four samples are averaged. The results are reported as:

P = maximum pressure;
L = time taken for the bubble to burst and is, thus, a measure of the extensibility of the dough;
W = the area under the curve, which is related to the strength of the dough; a high W indicates a strong flour.
P/L = dough strength and extensibility, *i.e.* the ratio of curve height to length. A low P/L is indicative of a very extensible but low strength dough.

5.2.2.1 Typical Results. Using the National Association of British and Irish Millers (NABIM) classifications, typical results for flour milled from British wheat varieties are shown below in Figures 1–4 and Table 1.

NABIM group 1 varieties, which are suitable for bread flour, give a strong and elastic dough.

NABIM group 2 varieties are used to make bread and baking flours. Most of these varieties have bread-making potential.

NABIM group 3 wheat is suitable for making biscuit and blending flours as it gives an extensible dough, and is also suitable for blending with strong wheats.

NABIM group 4 wheats are only suitable for animal food as they give flour that produces tough and inelastic doughs.

Flour from a NABIM group 1 wheat

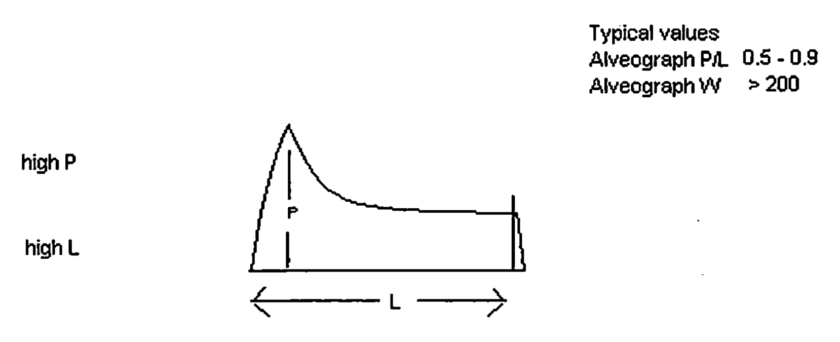

A bread flour giving a strong and elastic dough

Figure 1 *Chopin Alveograph results for flour from NABIM group 1 wheat*

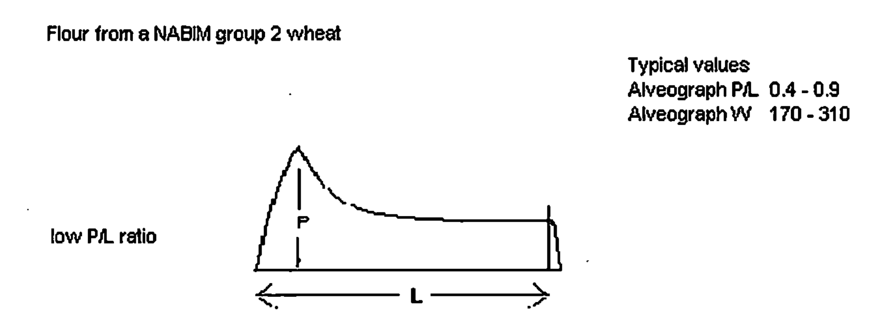

Suitable for bread and baking most varieties have good breadmaking potential

Figure 2 *Chopin Alveograph results for flour from NABIM group 2 wheat*

Although the Alveograph measures the same properties as the Brabender Extensograph there is no way of converting one set of results into another. Also, unlike the Extensograph, the dough is not kept for a long time.

5.2.3 Brabender Instruments

Brabender make a whole range of instruments for testing flour. These instruments are the standard ones in use in the UK, Germany and North America. The company has recently produced new versions of these instruments that use electronic measuring systems rather than the mechanical systems previously employed. The new models use the same name but with the suffix E. Thus, the new model Extensograph is the Extensograph-E.

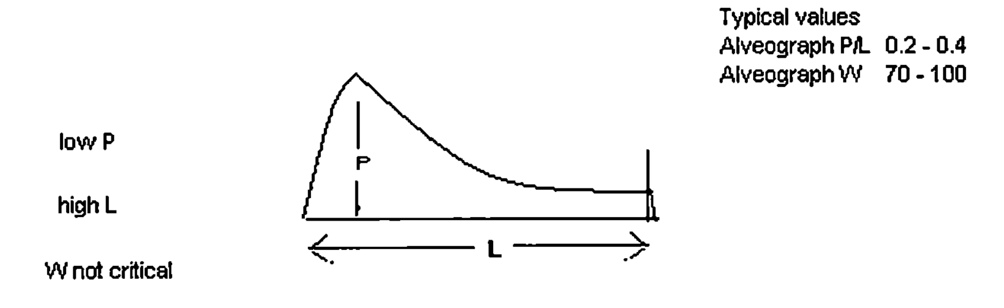

Figure 3 *Chopin Alveograph results for flour from NABIM group 3 wheat*

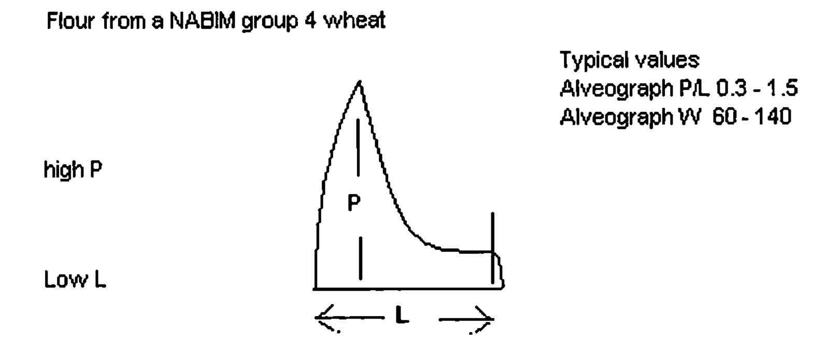

Figure 4 *Chopin Alveograph results for flour from NABIM group 4 wheat*

Table 1 *Typical results obtained with a Chopin Alveograph for flour milled from British wheat varieties (see text for NABIM group classifications)*

NABIM group	*P value*	*L value*	*Alveograph P/L*	*Alveograph W*
1	High	High	0.5–0.9	200
2	[a]	[a]	0.4–0.9	170–310
3	Low	High	0.2–0.4	70–100[b]
4	High	Low	0.3–1.5	60–140

[a] For group 2 a low P/L ratio is important.
[b] Value of W is not critical.

The new instruments can undertake a wider range of tests than the old instruments while being able to carry out the original tests. This is an important point since many of the original tests methods are international standards.

5.2.3.1 The Brabender Amylograph. This instrument gives information on the gelatinisation and amylase activity of flour (Figure 5). It works by measuring the change of viscosity of a flour water paste on heating at a uniform rate. The test corresponds to ICC standard No. 126/1, ISO 7973 and AACC standard no. 22-10. As the starch gelatinises the viscosity increases but the α-amylase present reduces the peak viscosity. The amylograph, like the Hagberg Falling Number only measures cereal α-amylases, any fungal α-amylase has no effect. The

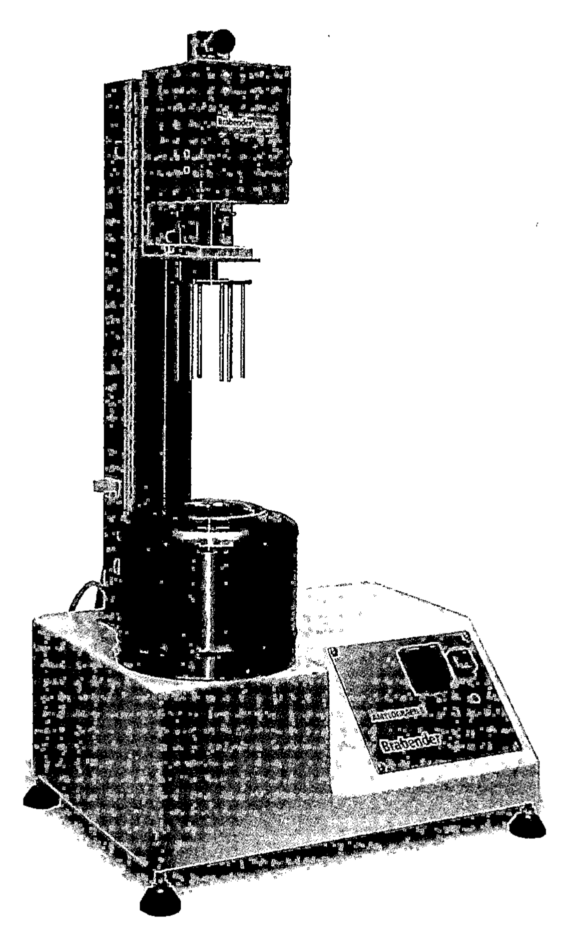

Figure 5 *A Brabender Amylograph-E*

heating rate is controlled at 1.5°C min^{-1} as this is the rate of increase of temperature inside a loaf during baking.

The instrument uses a heated bowl that is rotated. A stirrer, immersed in the bowl, is deflected according to the viscosity of the material in the bowl. The measurement system then plots this against time. In the Amylograph-E the data handling is electronic. This instrument can also carry out a rapid amylogram on a small quantity of flour by using a small mixing chamber and a higher mixing speed. The ability to work with small quantities is useful in plant breeding research.

5.2.3.2 The Brabender Farinograph and Farinograph-E. The Farinograph is used in a flour mill to measure the water absorption of flour. It does this by mixing flour and water to produce a dough of standard viscosity. In use a weighed quantity of flour is placed in the mixing chamber and water is run in from a specially calibrated burette. The viscosity used is 500 Brabender units in the UK and 600 Brabender units in North America. The mixer rotates at 63 rpm. The variation with time of the viscosity can be used to measure the mixing time and the stability of the dough.

The Farinograph-E (Figure 6) uses a PC to process the data from the Farinograph; this is a considerable improvement over measuring pieces of paper from a chart recorder. The software supplied runs under Windows® and can multi-task so that one PC can handle data from several instruments. This is particularly convenient as the Farinograph is used to prepare doughs for the Extensograph.

The water absorption measured by the Farinograph is normally reported as a percentage. The barbarous units of gallons per sack is sometimes encountered. A sack of flour contained 20 stones or 280 lbs of flour (approx. 127 kg). This unit was used when mixes were referred to as a certain number of sacks of flour and the water added was measured in gallons.

Farinograph doughs only contain flour and water, so the water absorption obtained is a theoretical value. The value obtained from the same batch of flour in a bread dough will always be lower. When the Farinograph is used to prepare doughs for the Extensograph, the doughs do contain flour, salt and water.

In a flour mill the Farinograph is used as a control instrument since the water absorption of flour can be increased or reduced by adjusting the mill, to deliver more or less starch damage. If the water absorption is too low, the pressure on the reduction rollers of the mill will be increased, thereby increasing the starch damage. Similarly, if the water absorption and hence the starch damage is too high the pressure must be

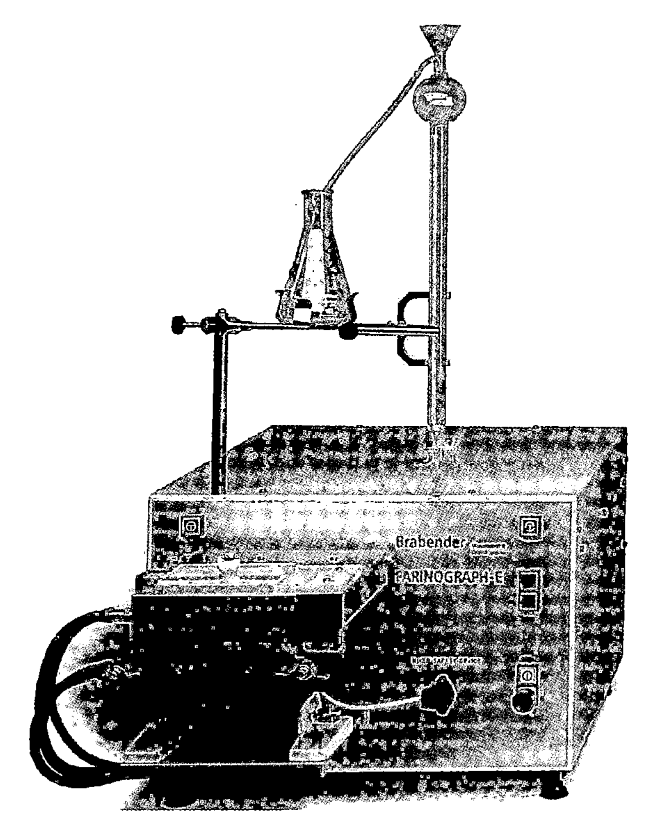

Figure 6 *A Brabender Farinograph-E*

reduced. Water absorption is monitored throughout the run as the smallest variations in the wheat or the mill cause the water absorption to move out of specification.

In a bakery the Farinograph can be used to test whether the flour is in specification or not. It would be possible to add more or less water, depending on a measured water absorption.

The major use of the Farinograph in a bakery laboratory is to prepare doughs for the Extensograph.

5.2.3.3 The Brabender Extensograph and Extensograph-E. The Extensograph (Figure 7) again has different uses in the milling industry to the ones it has in a bakery. In the flour mill one of the major uses of the Extensograph is to monitor the effectiveness of the flour treatments being used. This would be a major concern in a bread flour mill when the new season's wheat starts to arrive as a treatment regime that was

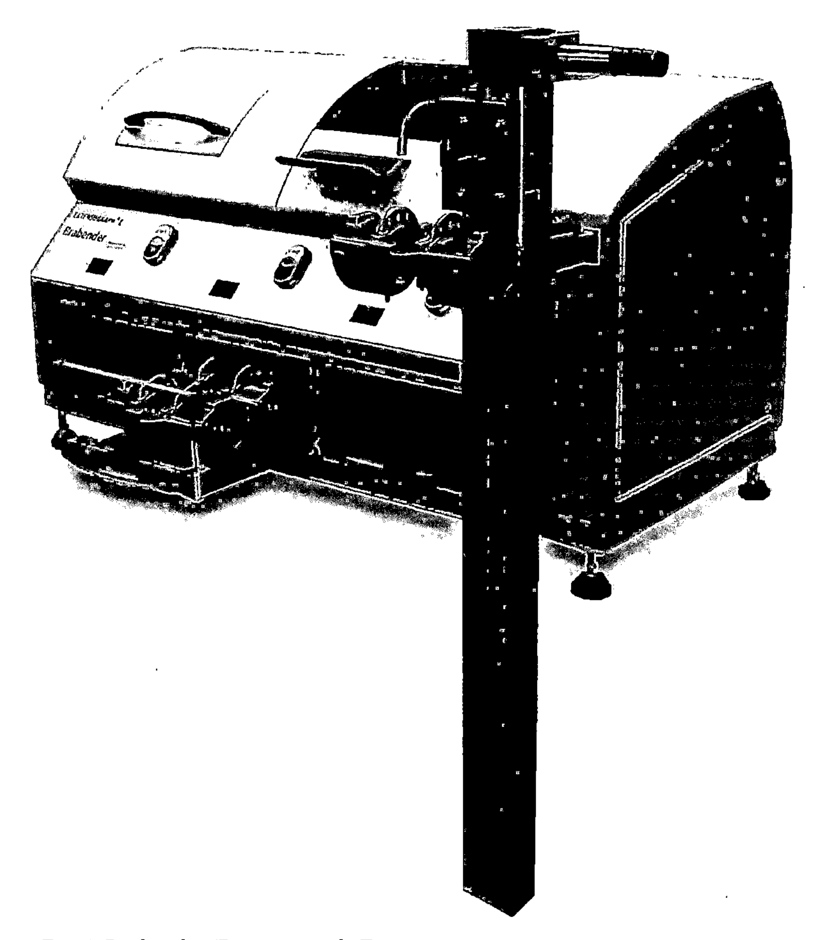

Figure 7 *A Brabender Extensograph-E*

satisfactory for one year's harvest might be totally unsuitable for the following year.

The Extensograph works by taking chubs of a dough made from distilled water, flour and salt. The dough is made in the Farinograph. The dough is then machined on the Extensograph and made into a number of chubs. Some of these are stored in a temperature controlled cabinet for later testing. Three chubs are then tested by stretching them with a hook in a controlled way, one after the other, and the results are averaged. The force required to stretch the dough and the amount that the dough stretches is recorded. The testing regime is repeated after both 1.5 and 3 hours. The output is referred to as an extensogram (Figure 8). The resistance is defined as the height 5 cm down the chart while the

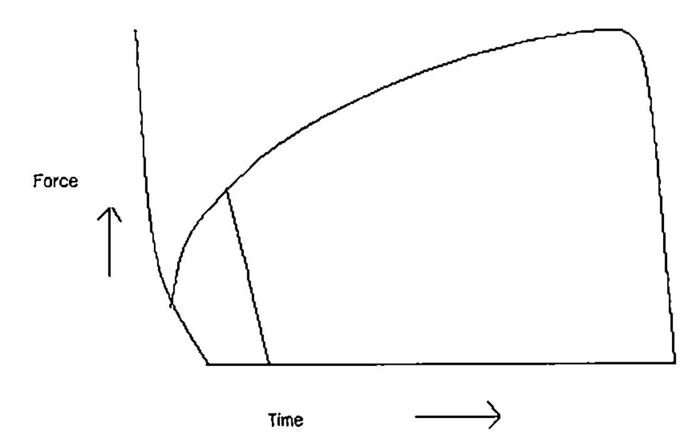

Figure 8 *A general extensogram*

maximum is the maximum height reached, *i.e.* the point at which the dough broke. The extensibility is the length that the dough was stretched before it broke.

A complete set of measurements from the extensogram will give the energy, *i.e.* the area under the curve, the resistance to extension, the extensibility and the maximum, *i.e.* the deflection when the dough broke. The ratio of the extensibility to resistance and the ratio of extensibility to maximum are calculated. The Extensograph-E is set up to calculate these values directly.

Experienced Extensograph users learn the characteristic shapes of the extensogram typical for flour suitable for particular uses (Figures 9–11). Bread flours need both resistance and some extensibility while in biscuit flours a low resistance and a high extensibility are needed.

The results of the 1.5- and 3-hour samples give information on the stability of the dough, which is of importance in any long process. Some Extensograph users do not bother with the long-standing times as their products are made by a rapid process.

The Extensograph is used in several international standards, specifically ICC standard no. 114/1, ISO 5530-2 and AACC standard no. 54-10. These standards are important in the international trade in flour; a buying specification written in terms of Extensograph testing to method ISO 5530-2 would not necessarily give the same answers as AACC 54-10.

Typical Values from the Farinograph and Extensograph. Different sorts of flour might have values such as those shown in Table 2.

5.2.3.4 An Example of the use of the Extensograph. The Extensograph is very useful in sorting out problems where several flour samples are successful in an application while others of similar analytical

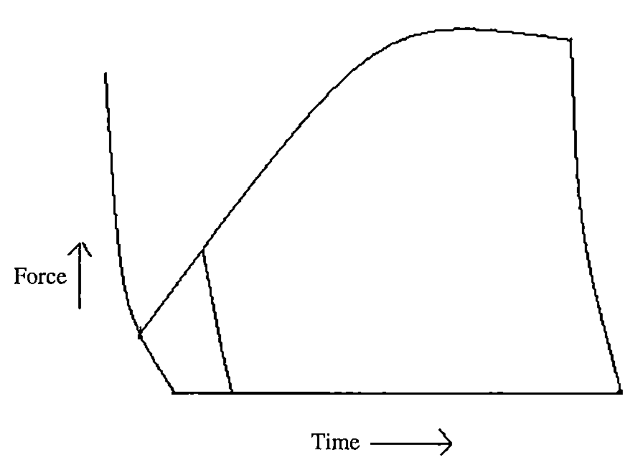

Figure 9 *Extensogram of very strong flour*

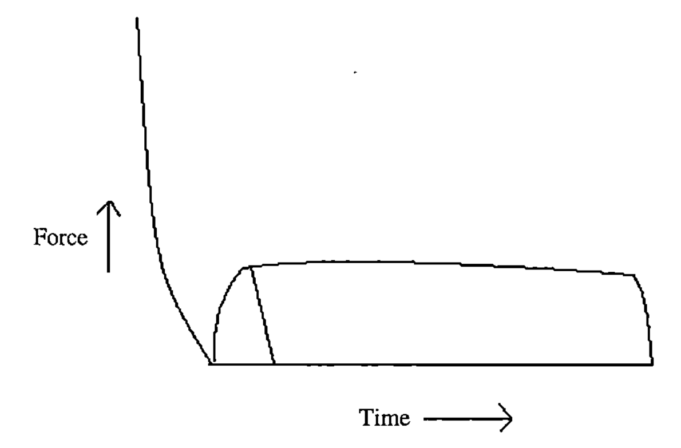

Figure 10 *Extensogram of medium flour*

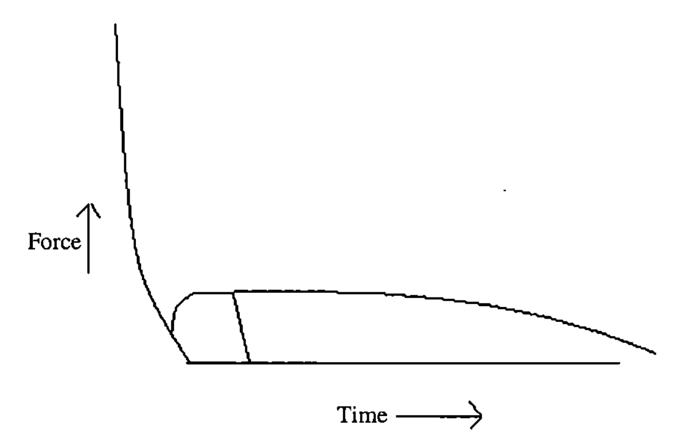

Figure 11 *Extensogram of weak flour*

Table 2 *Typical Farinograph and Extensograph results*

	Farinograph values			*Extensograph values*	
Type of flour	*Water absorption (%)*	*Development time (min)*	*Stability (min)*	*Resistance (BU)*	*Area (cm^3)*
Top quality bread flour	62–64	5	15	300	130
Baker's grade bread flour	60–62	5–7	9–12	250–350	120–140
Soft wheat flour	58–60	2–4	3–5	150–200	60–80

properties are not. This is more likely to happen when the product is not bread. The properties needed for a successful loaf are well understood while those for other products are less well understood.

In the example a plant bakery was making puff pastry from the sort of all-EU bread flour normally used in a CBP bread plant. The puff pastry was used to make pies and sausage rolls. When the pie casings had been cut the scraps were re-rolled and recycled.

Some flour samples caused the pastry to snap when being rolled while others of similar composition worked. A treated bread flour is less than ideal in this application since gluten development is undesirable but the rolling and re-rolling causes mechanical development of the gluten. Extensograph testing revealed that this product needed a minimum extensibility from the flour. The solution was to use an untreated flour with a weaker grist. Incidentally, all the flours worked perfectly well in CBP bread plants.

5.2.3.5 The Brabender Maturograph. This instrument is used to measure the volume change that occurs when a dough is proved. Unlike the Farinograph and Extensograph it uses a complete bread dough. The dough is placed in a temperature controlled chamber and the volume change with time is recorded. The instrument gives the final proving time, the proving stability, the elasticity and the dough level.

5.2.3.6 The Brabender Oven Rise Recorder. This instrument is used to measure the amount that a dough rises in the oven. A dough sample is heated by being placed in a temperature-controlled oil bath. The instrument measures the buoyancy of the dough. As the dough expands under the influence of heat the up force increases according to Archimedes principle. The output of the instrument is the dough volume, the baking volume, the oven rise, the final rise and any peaks caused by gas escaping.

5.2.4 The Mixograph

This instrument is not produced by Brabender. It is an American instrument that was designed in 1933 by Professor Swanson. In use a dough is mixed in a high speed pin mixer to give a mixing curve that is used to measure flour strength. The time taken to discriminate between flours giving tough stable doughs is about half that taken with a Farinograph. This sort of flour is not often encountered in the UK, which explains why this instrument is little used in the UK.

5.2.5 The Grade Colour

Unlike the other tests in this section the grade colour is not a performance test but is a test of the whiteness of flour. The Kent-Jones and Martin colour grade is measured by comparing the reflecting power of a dispersion of flour in water with a standard reference surface. A few patent flours give a negative colour grade, which merely indicates that they are whiter than the reference.

An example of one of these instruments would be a Lovibond flour colour grader series 4. In use the following regime is followed:

A flour sample is weighed then mixed with water for a standard time and poured into a glass cuvette. The cuvette is inserted into the instrument, which moves into the measurement position and takes the reflectance reading. The results appear on a LED display and are printed out. The instrument is calibrated with an internal ceramic tile and standardised using a national standard flour. The normal range of flour grades is –5 to +18.

5.2.6 The Sodium Dodecyl Sulfate (SDS) Test

This test, which has been adapted as BS 4317 AACC 151 and ICC 56-70, is used to determine the bread-making quality of wheat or flour. It is really an acceptance test for a flour mill rather than one for the bakery laboratory.

The flour or ground wheat is shaken with water to which the sodium dodecyl sulfate is added. After a series of timed inversions the cylinder containing the sample is allowed to stand for 20 min. The height of the sediment is then read and recorded.

5.2.7 The Cookie Flour Test

There is an American test for biscuit flour that involves producing standard cookies baked on a standard aluminium plate. The diameter of the finished product is measured. The greater the diameter the higher the score. This test is of course a measure of dough extensibility, which is the crucial property for biscuit flours.

CHAPTER 6

Baking Machinery

6.1 INTRODUCTION

There are several operations in making baked products and there are the machines that are used to carry them out. While there has been a revolution in the baking of bread, the fundamental processes of baking products have not changed – some of them are merely carried out in a different way. There are bakeries all over the world, including the third world, and these bakeries do things in different ways.

Progress in bakery machinery has been aimed at replacing manual effort with machinery, reducing the amount of labour involved and speeding up the processing. The speed of this change has been very rapid. There are bakers still alive in the UK who have made bread doughs by hand in a mixing trough.

The Chorleywood process produced a considerable change in the way that many plant bakeries work. Even the Chorleywood process has evolved since potassium bromate has been prohibited and the process had to accommodate the use of other flour treatments.

The ADD process offered the small baker an opportunity to make no time doughs in existing mixers. Unfortunately, the removal of potassium bromate from the permitted list made this process unworkable.

Spiral mixers offered the small baker the opportunity to produce no time or short time doughs without the use of specialised improvers (Figure 1). These machines have helped to keep small bakers in business.

Bread of course is not the only bakery product. While other products have not undergone a revolutionary change similar to the one that has happened in bread making there have been continued incremental improvements. These changes have been directed at saving labour and time. Energy saving has been a major concern of oven designers for some time as it is the most energy intensive part of any bakery process. Otherwise energy saving has not been a particular concern in bakery equipment.

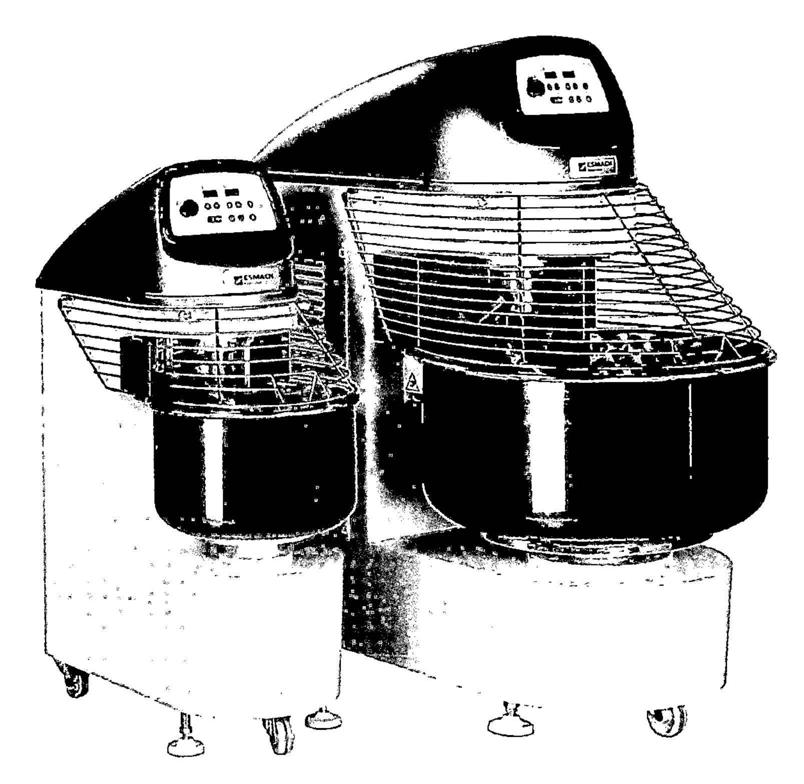

Figure 1 *A spiral mixer*

One development that has affected bakery equipment is the microprocessor revolution. The availability of cheap computing power has seen mechanical controls superseded by electromechanical controls to be replaced by electronic controls. It is now possible for an entire bakery to be controlled by one ruggedised microcomputer. This is a considerable cost reduction compared with the mini computer that would have been required previously.

There is one piece of equipment that approaches the process engineer's dream of a machine that takes in raw materials at one end and finished product emerges from the far end. That machine is the high pressure extruder. Unfortunately, while high pressure extrusion can make crispbread it has a limited application in the bakery sector.

6.2 MIXING

6.2.1 Bread Dough Mixers

A bread dough mixer has to carry out two functions, to make the dough and to knead it. The early mechanical mixers used a two-arm system, as seen in mixers such as the Artofex. These mixers were designed to imitate the action of hand mixing. There was a fear of applying too much energy to the dough so a gentle action was preferred. Considering the amount of energy applied to bread doughs in the Chorleywood process this is now surprising.

The big change in bread mixers was the introduction of the Tweedy mixer with the Chorleywood bread process. The Tweedy mixer was developed not for the Chorleywood process but originally for non-food uses. The original Tweedy mixers used in the Chorleywood process simply applied a vacuum. The modern APV Tweedy mixer first applies pressure, causing more air (particularly oxygen) to dissolve in the dough. Then a vacuum is applied that causes the air to form larger bubbles. The extra oxygen made available as a consequence of the pressure stage means more effective oxidation, which requires less ascorbic acid, which is an expensive ingredient. An increased yield is possible because the dough is machinable at a higher water content. The crumb colour is also improved. The upgrade to Tweedy mixers can be applied to older machines.

Modern Tweedy mixers can be supplied with electronic weighing apparatus to control the flour:water ratio. The entire system can be controlled on a recipe basis and can reconcile the use of ingredients. The system can maintain accurate dough temperatures by blending in cold water and accurate control of the energy used.

The current producer of these machines is Baker Perkins Ltd, which has (as of April 2006) regained its independence and has separated from Invensys plc.

While it is possible to demonstrate that it is possible to use too much energy in a Chorleywood plant, overmixed doughs are not a common bakery fault. Doughs that have failed to develop through under mixing are much more common.

Electronics play a considerable part in the energy control systems used in Chorleywood plant mixers. It is possible to measure the energy input to a direct current motor simply by measuring the voltage and current and multiplying them. This does not work on alternating current motors since current and voltage are themselves fluctuating and an AC motor can be delivering considerable energy with a low current but a negative phase angle. Those requiring further information on this issue are directed to a textbook on AC theory.

The next development in the bread dough mixer story was the spiral mixer. These mixers are so-called because they use a spiral element to mix the dough. They provide a more gentle mix than a Tweedy mixer, but a spiral mixer can usually deliver in under 15 min a developed dough. This should be compared with the 5 min that a Tweedy mixer would take.

Some spiral mixers have only a single speed while others have several speeds. These mixers often have a reverse setting designed to, for example, fold dried fruit into bun doughs. This setting is intended to mix with the absolute minimum of energy. It is not too surprising that when complaints of poor performance arise it has been found that, accidentally, the reverse setting has been engaged when a dough should be developing and does not!

6.2.2 Biscuit Dough Mixers

Biscuit doughs have different needs in mixing. While all types need the dough to be mixed, yeast-raised crackers need the gluten to be developed, semi-sweet biscuits need some dough development, and short doughs need to avoid development.

Some biscuits are mixed in continuous mixers to feed continuous shaping and baking installations. The problem with continuous mixers is that they tend to cause problems at start up and need expensive continuous metering apparatus to control the addition of ingredients. If the is a hold up elsewhere in the plant the mixer must be either turned off or the dough must be directed to a holding vessel. Turning off the mixer will lead to the problem of restarting it. A holding bank can only be a temperary solution at best.

Batch mixers can cope with a plant that is shaping and baking continuously because mixing can be done much more quickly than shaping or baking. The moulding or cutting machines normally have a feed hopper, which needs to be topped up with dough from time to time.

Both horizontal and vertical mixers are used for biscuit doughs and both have their advantages and disadvantages. Vertical mixers usually have a removable bowl that can be wheeled across the shop floor to move dough. Spare bowls can be held ready to use.

The horizontal mixers normally used consist of a "w" shaped mixing bowl with two "Z" blades (Figure 2). These mixers do not have a removable bowl so the dough has to be tipped out. The mixers can deliver large amounts of energy at low speeds.

6.2.3 Cake Mixers

Cake mixers are usually of the planetary type (Figure 3). In some ways they are a scaled up version of a domestic table mixer. Some cake mixing involves whipping eggs so there is a requirement to beat at high speed

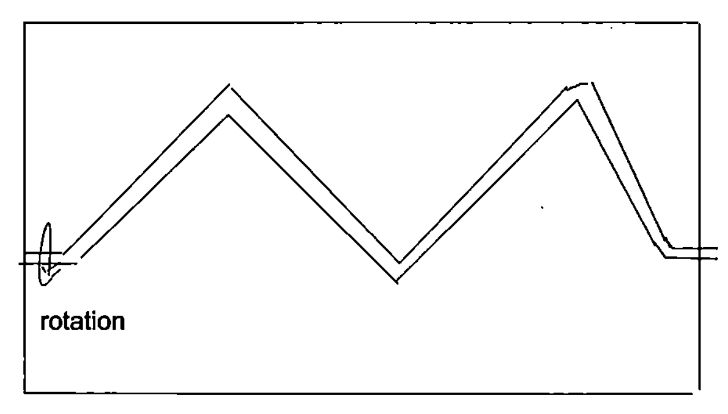

Figure 2 *Schematic of a "Z" blade mixer*

Figure 3 *A planetary mixer, used to make cakes, for example*

and incorporate air. Cakes are one of the systems where dough development is not wanted.

6.2.4 Pastry Mixers

In mixing pastry the important thing is not to cause gluten development. Any mixer can produce pastry. Small bakers tend to use planetary mixers for this.

6.3 MEASURING AND WEIGHING INGREDIENTS

Modern bakeries are equipped to precisely measure the quantities of materials being used. The water going into a batch is measured by a water meter (Figure 4). This is a far cry from using a measuring bucket. The old fashioned way of reporting the water absorption of flour as so many gallons per sack made some sense when the mix would be a sack of flour and a number of gallon buckets of water.

Where bagged flour is being used the flour and recipes are often worked out as a quantity of bags or sacks. This avoids any need to weigh flour in the bakery as the miller has already weighed it into the bags. The use of bulk flour means that flour will have to be weighed. In large CBP plants all the ingredients are metered into the system. The most modern systems can actually work out what needs to be added once a recipe is selected (Figure 5).

In smaller set ups or where small batches of product are made from a bulk supply of flour the flour needs to be weighed. Now this can be done using ordinary scales but there are systems that make the baker's life easier. It is possible to have a system with a load cell under the mixing bowl that allows the flour to be weighed in directly. The water can fed in from a water meter and any other ingredients can be weighed in.

While industrial weighers are rugged and can survive in an environment with heat and flour dust, a weighing system of laboratory precision can not operate in such an environment. Even if very small quantities were needed there would be the problem of mixing them uniformly throughout the product. Where potassium bromate is still legal it is the practice to add this powerful oxidising agent to the flour, having previously diluted it with an inert material. The consequences of mixing any powerful oxidising agent with an organic material can literally be explosive. There are then several reasons why minor quantities of ingredients need to be pre-dispersed in some other medium before addition to the dough. This is the problem that compound improvers address. All the minor ingredients for a product are dispersed in either another ingredient or an inert material. With bread this could include soy flour, ascorbic acid and fungal α-amylase. When adding L-cysteine hydrochloride to puff pastry as a dough relaxant the active ingredient is pre-dispersed in either soy flour or heat treated wheat flour. The soy flour can itself add useful colour to the pastry through the Maillard reaction, while the wheat flour has been heat treated to inactivate the enzymes present.

These compound improvers can then be added in easily measurable quantities. This is an advantage not only on a small scale but also in

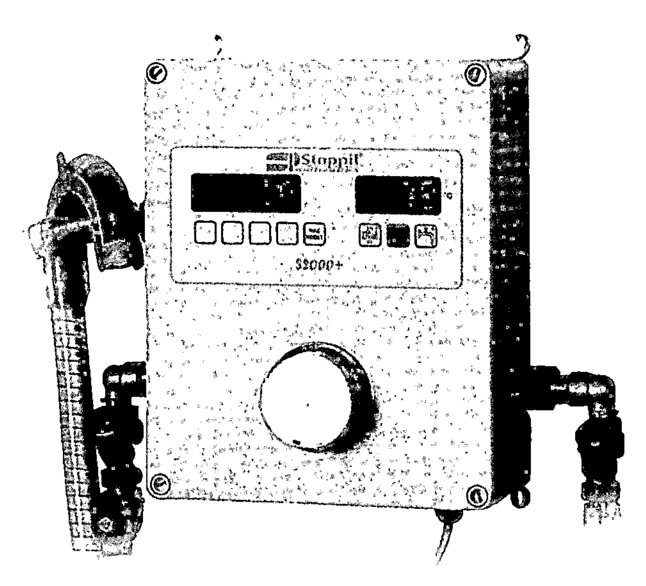

Figure 4 *Water meter used to measure water in bread dough*

Figure 5 *A one-man operated bread plant*

large plants. If several ingredients are added as a compound improver they can be added by one feeder. Additional feeders and their control gear are expensive.

6.4 PROVING AND RETARDING

Proving is the practice of holding yeast based products under controlled temperature and humidity conditions. The temperature is designed to maximise yeast growth while the humidity is normally high to prevent the dough forming a dry skin. The temperature used does vary, depending on the product. The prover is then the cabinet that is set up to do this. Proving is sometime referred to as fermentation, which is what is happening.

Retarding is holding a dough at a low temperature, typically around a refrigerator temperature, to hold up the fermentation process. Under these conditions the activity of the yeast is slowed but the activity of the flour amylase enzymes is reduced much further. Retarding can prevent a dough "going rotten" as the action of the amylase breaks up so much starch that the dough becomes sticky and unhandleable. The likelihood of this happening is much greater with doughs based on flours with a higher English wheat content than with those based on Canadian wheat. The ability of a dough to stand these conditions is referred to as its tolerance.

One use of the retarder is in emergencies where a breakdown somewhere further down the line causes a hold up. In these circumstances the retarder is regarded as a life saver.

In some products a spell in the retarder can be part of the process. This can be the case with some pizza doughs.

The retarder can be used to ease the flow of work through the bakery. This applies to both large and small bakeries. In a one-man bakery the retarder can be used to make production easier. As an example, if the first dough made is sent to the prover a second and subsequent doughs can be sent to the retarder. When the first dough has proved it can be shaped scaled and panned and sent for a second proof. The second dough is moved from the retarder to the prover so it can prove while the first dough is being shaped, scaled and panned. The use of the prover and retarder then ease the work flow through the bakery.

6.5 SHAPING AND PANNING

Most British bread is shaped by placing it in a tin, a process known as panning. In large bakeries panning is automated.

Some other breads are shaped by machine or hand. Machines exist to shape baguettes or rolls automatically.

Cakes are normally shaped by making them in tins. As cake batters flow the use of a tin is more or less essential.

Biscuits can be shaped by cutting, either rotating or vertical, wire cutting or rotary moulding. These processes are covered in Chapter 8 (Section 8.4 on biscuits).

Pastry is normally shaped by rolling and cutting. The pieces are than shaped as required for pies, sausage rolls or whatever other product is needed.

6.6 SCALING

Scaling is the process of weighing the dough out. This is important since where bread is sold by weight it has been an offence to sell short weight since the middle ages. At the time of writing, UK legislation requires that all bread above 300 g in weight is sold in prescribed quantities: 400 g, 800 g and 1200 g. These of course are near to 1 lb, 2 lb and 2.5 lb in imperial measure. The machinery in the bakery has to ensure that under-weight loaves do not leave the bakery. This is usually achieved with a check weigher.

6.7 BAKING

This is the final process in the system. The product enters the oven and heats up either by convection, conduction or radiation, or possibly a combination. In yeast-raised products the final expansion takes place, known as oven spring. Excessive oven spring is a sign of insufficient final proving. A moderate amount of oven spring is a good sign.

In yeast-raised products the final expansion comes from the carbon dioxide from the yeast, expansion of air and water turning into water vapour.

In chemically leavened products the same things occur except that the carbon dioxide is chemically produced. The rate and temperature at which the carbon dioxide is produced depends on the choice of chemical leavening agents.

In some sponge cakes the expansion is caused purely by the expansion of the air and the water held in a foam that is stabilised by egg. In other sponges chemical leavening agents are also present.

Once the final expansion has occurred the proteins start to denature and the starch gelatinises. The proteins also undergo Maillard reactions, particularly at the outside of the product, which is the hottest part.

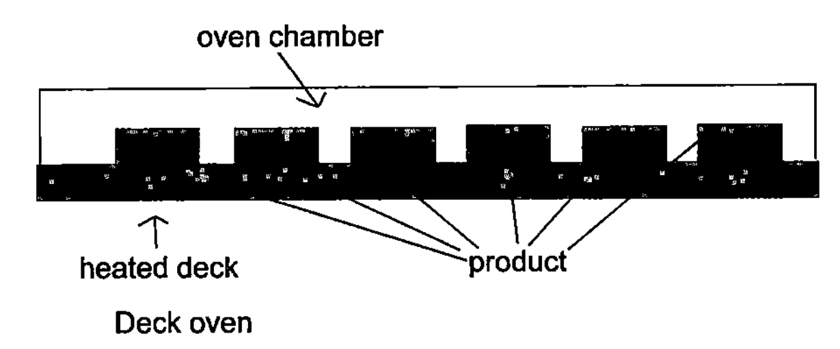

Figure 6 *Schematic principle of a deck oven*

While all of the above are happening water is being lost from the product. Towards the end of the process water loss can become the only thing that happens, particularly in biscuits.

Ovens can be heated by gas, oil, coal, coke, wood or steam. If an oven is steam heated there has to be a steam boiler elsewhere to generate the steam. This should not be confused with steam baking where steam is introduced into the oven to give a particularly crisp crust. A low technology way of doing this is to put a tray of water in the bottom of the oven.

While there are small bakers who are using old ovens that consist merely of a heated box most modern ovens fall into one of three classes: They are either deck ovens, rack ovens or travelling ovens. Travelling ovens are also known as tunnel ovens.

Deck ovens resemble a cupboard with several chambers (Figures 6 and 7). The floor of each compartment is heated so heat transfer is mainly by conduction. These ovens are used among others by hot bread shops and those bakeries selling pies, pizzas, sausage rolls and similar items.

Rack ovens have central rack which is rotated around a vertical pivot (Figures 8 and 9). The rack accepts trays of products. The rotating rack evens out the flow of heat to the products. As the rack rotates hot air is blown over the products so that a very even heat distribution is obtained. Some rack ovens are equipped to blow steam over the product, either to give crisp baguettes or to steam products like Christmas puddings.

Travelling ovens can bake continuously since they consist of a conveyor that is either a steel band or a wire mesh that travels through the oven (Figure 10). The product goes into the oven raw and emerges fully cooked. Tunnel ovens are normally set up with several zones whose temperature can be controlled independently. Thus they can be arranged to cook on a declining heat. The ovens normally work by blowing heated air over the product, *i.e.* they work by convection.

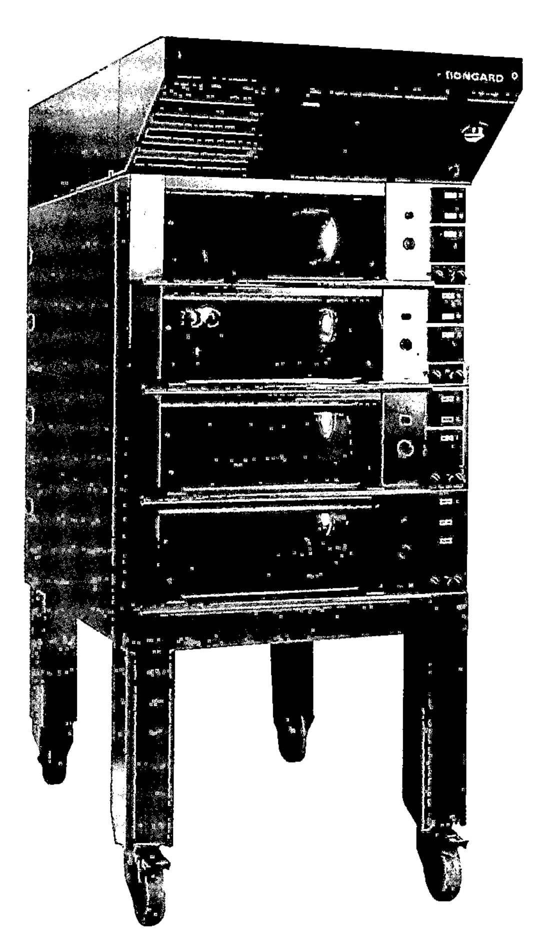

Figure 7 *A deck oven*

6.8 EXTRUSION

Extrusion is the first really new technology to arrive in the food industry for many years. Tunnel ovens are a development of an oven while the sort of mixer used for the CBP process is a development of a mixer, albeit one that can apply a great deal of energy in a short time.

An extruder can, under suitable conditions and with the right ingredients, mix, cook, knead, shear, shape and form. Claims are made that an extruder saves capital cost since the capital cost of an extruder is less than that of the other equipment needed to perform all these operations. This may be true but is not necessarily the way that an existing bakery would view things since they probably already have the traditional equipment and need to purchase an extruder.

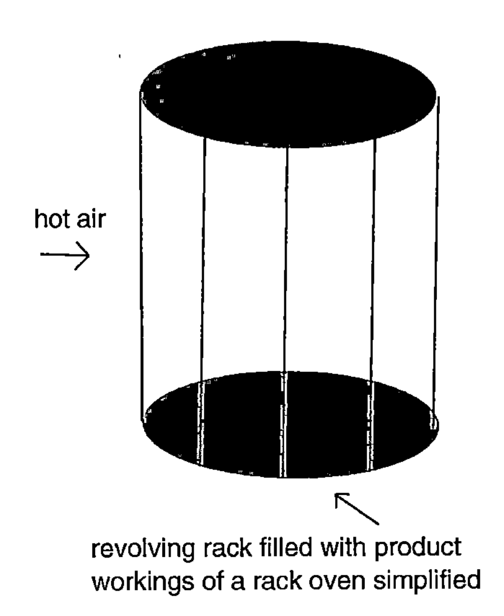

Figure 8 *Simplified schematic of a rack oven*

Figure 9 *Product being removed a rack oven*

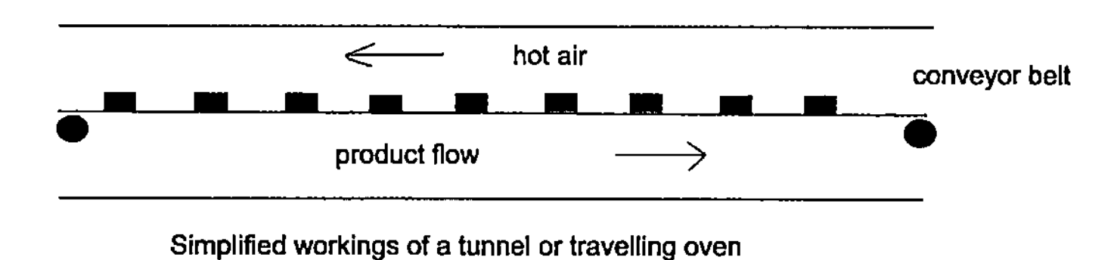

Figure 10 *Simplified scheme of a travelling oven*

6.8.1 Classification of Extruders

Extruders are normally classified by their method of construction, *i.e.* twin screw or single screw and the operating conditions, *i.e.* cold extrusion or extrusion cooking. Cold extrusion is used to make liquorice and pasta among other products and so is outside the scope of this work. Extrusion cooking, which is defined as heating the product above 100 °C, has many uses.

6.8.2 Extrusion Cooking

Extrusion cooking comes into the scope of this work because products like crisp breads can be made on an extruder. Instead of mixing a dough in a mixer and shaping it and baking it the flour and other ingredients are fed into the barrel of the mixer. The rotating elements and the stationary elements combine to mix and shear the ingredients. As the mixture is forced through restrictions in the barrel the pressure rises. The intense work, and possible external heat, heat the mixture. As the pressure is above atmospheric the boiling point rises. Under these conditions very rapid cooking occurs. The product is then forced through as slit. The water then flashes to steam, expanding and further cooking the product. The latent heat of evaporation of the steam helps to cool the product.

CHAPTER 7

Bread Making

7.1 THE CHEMISTRY OF DOUGH DEVELOPMENT

Dough development is a fundamental process in bread making, without it there is just a paste of flour, water and the other ingredients. If the bread is to expand and form a proper cell structure then this change must take place. While it is quite easy to test for dough development by prodding some dough with a thumb it is more complicated at a chemical level.

Rheologically a developed dough is viscoelastic. This can be verified simply – as the dough is worked it heats up, indicating that it has viscous behaviour. The elastic behaviour can be demonstrated by deforming the dough and watching it spring back.

At a molecular level it is believed that bread improvers oxidise the cysteine sulfhydryl or thiol (–SH) groups present in wheat gluten. This then prevents the thiol groups indulging in exchange reactions with the disulfide (–S–S–) bonds. Another possibility is that the oxidation of the –SH groups may lead to the development of new –S–S– bonds, thus cross-linking proteins. Dough development then involves the breaking of some bonds and the formation of new ones. These bonds hold the protein in its original configuration. After these bonds have been broken the proteins can join up to form a three-dimensional matrix. It appears that the same process occur in a wholemeal dough, where there are no flour treatments.

The scientific understanding of dough development is, admittedly, less than perfect. This is not too surprising given the difficulties of obtaining structural information from a complicated mixture like a flour dough.

In bulk fermentation and sponge batter processes the bonds are broken by the action of enzymes and flour improvers. In the Chorley-wood process the bonds are broken by intense mechanical input and the action of the improver. Similarly, in an ADD process the effect is produced by the reducing action of the cysteine followed by the slow-acting oxidation of the potassium bromate accompanied by mild mechanical action. There may be some insight into the process in that attempts to replace potassium bromate with ascorbic acid or azodicarbonamide, which act more quickly, have failed.

7.2 THE MAKING OF BREAD

There are a whole range of methods of making bread; however, it is possible to classify them (Table 1 and Figure 1). Craft bakers tend to be individualists and often have their own variants of methods.

7.2.1 Unleavened Bread

This has to be the simplest method. An example of an unleavened bread is the chapatti, which originates on the Indian sub-continent.

7.2.1.1 Method

Mixing. The flour is made into a dough with water and possibly some salt.

Dough Development. The dough is normally left for around 30 min before shaping although some chapatti makers make a large batch of dough leaving it in the refrigerator until needed.

While the dough is being rested the amylase enzymes in the flour will attack the starch and soften the dough.

Table 1 *Methods of making bread*

Method	Bulk fermentation	Sponge and batter	Continuous mixer	Chorleywood
Mixing	Conventional	Conventional	High speed	Tweedy high speed mixer or similar
Loaf making	Conventional	Conventional	Divider/ panner	Conventional
Automation	Low	Medium	High	High
Fermentation loss	Medium	Low	Medium	High
Tolerance	Medium	High	Low	Low
Total time (hours)	3–4	5–6	3–5	2
Variations	No time dough	Sponge dough or flour brew		Conventional loaf making

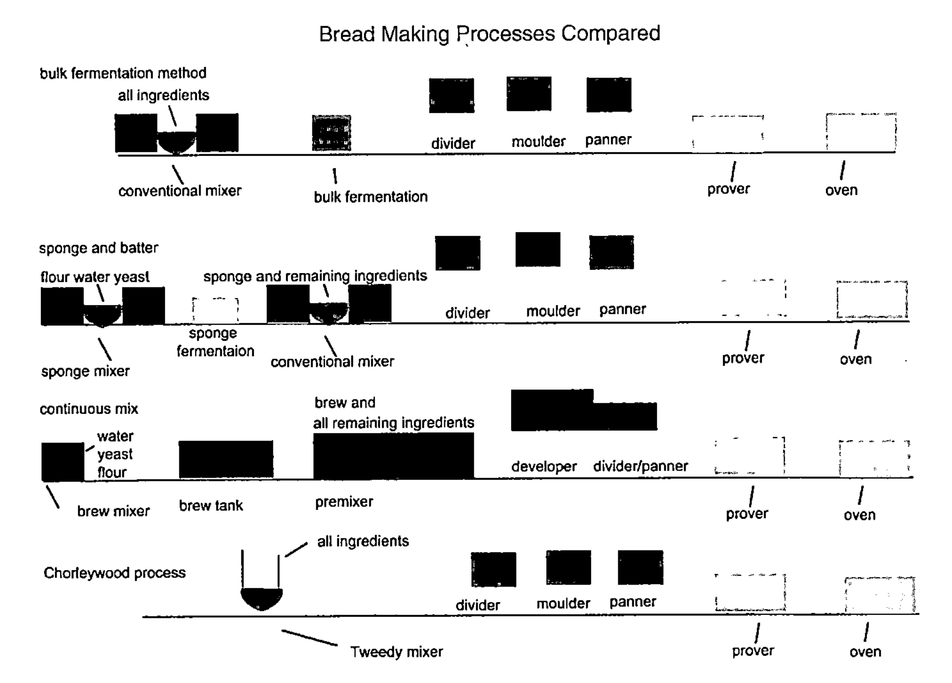

Figure 1 *Comparison of bread-making processes*

Shaping. The dough is then rolled out to produce circular pieces.

Cooking. The circular pieces are cooked on a hot griddle and, possibly, then being exposed to radiant heat. The product has to be eaten straight away. Chapattis are sometimes coated with ghee or melted butter.

7.2.2 Sour Dough Bread

Presumably, at some point in history a batch of unleavened bread dough became contaminated with wild yeast and the first leavened bread was made. The only way in which such a system can be kept going is to keep back a portion of the old dough and add more flour and water to keep the fermentation going. In such a system various side-reactions occur and fatty acids start to be produced, hence the name. The acids give the dough some protection from contamination with undesirable moulds, which would be an advantage in a primitive society.

Sour dough bread is made all over the world in both primitive and advanced societies. In primitive societies there is no alternative, but in advanced societies sour dough bread is made for its special flavour. Sour dough is also added to bread made by other processes to enhance the flavour.

Sour dough bread has a range of flavours that are not present in other bread; also it keeps relatively well. The keeping properties, no doubt, originate from some of the products of the side-reactions that have taken place in the dough. There is no need to add propionic acid as a mould inhibitor as some is likely to be present naturally. Quite possibly, some of the substances produced by the fermentation would not be permitted as additives!

7.2.2.1 Method

Mixing. Sufficient flour, water and possibly salt are mixed with a starter of old dough to produce a new dough. Commercial starters for sour dough are available.

Dough Development. The dough is kneaded either by hand or mechanically.

7.2.3 Bulk Fermentation

This is the method used by most traditional small bakers in the UK. It is also used by most domestic bread makers. It has been used in plant bakeries but this is not now common in the UK. It is a common method in the USA, while it is still used in some Scottish plant bakeries. Figure 2 shows a white loaf made by bulk fermentation (a close up is shown in Figure 3). Figures 4 and 5 show similar views of a wholemeal loaf.

7.2.3.1 Method

Mixing. The flour, water, yeast and salt are mixed together with any other ingredients. Other possible ingredients include fat, enzyme active soy flour and flour improver.

Dough Development. The dough is then kneaded either by hand or mechanically. A considerable amount of energy normally goes into the dough at this part of the process. If the dough is hand kneaded the energy input rapidly becomes apparent to the kneader. While the dough is being kneaded its rheology changes to become viscoelastic. This indicates the formation of a three-dimensional network in the dough. Old fashioned craft bakers would use the feel of the dough as an indicator that it had been kneaded sufficiently.

Proving. In this step the dough is left to rise. The yeast multiplies and produces carbon dioxide gas and alcohol from sugars. The sugars are

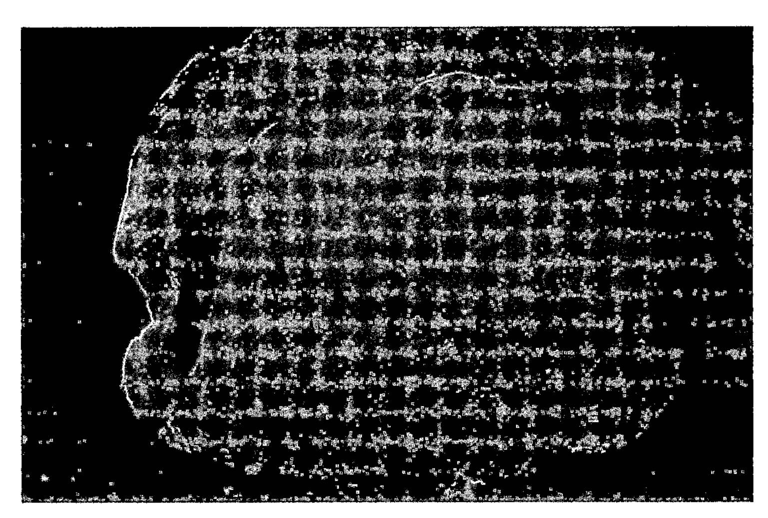

Figure 2 *A white loaf produced by the bulk fermentation method (note the uneven bubble size)*

Figure 3 *Close up view of a white loaf produced by the bulk fermentation method*

either those added initially or liberated by the yeast from the flour. Any sucrose will have been inverted to fructose and dextrose by invertase from the yeast.

One of the changes that happens during the dough development stage is that the dough traps the carbon dioxide.

Knocking Back. The risen dough is then re-kneaded to remove large gas bubbles.

Figure 4 *A wholemeal loaf produced by the bulk fermentation method (note the uneven bubble size)*

Figure 5 *Close up view of a wholemeal loaf produced by the bulk fermentation method*

7.2.4 Sponge Batter or Sponge Dough or Flour Brew

This method is used in several countries, including France and the USA. The characteristic difference with other methods is that a portion of the flour, *e.g.* a third, is fermented with the water the yeast and any added sugars.

Some bakers have used a stronger flour, *i.e.* one with lower amylase levels, for the fermented part of the process. Obviously this is the part of the operation where a flour with excessive amylase levels, *i.e.* too low a falling number, is most exposed.

The sponge batter method is not much used in the UK, except in some Scottish plant bakeries.

7.2.5 Chorleywood Bread Process

This process takes its name from the Flour Milling and Baking Research Association that was situated at Chorleywood in Hertfordshire. The Association has since merged with the Campden Research association and has moved to Chipping Campden in Gloucestershire.

The fundamental difference between this process and bulk fermentation or sponge batter processes is that the dough development is achieved by a combination of high mechanical energy and chemical action.

Curiously, while a biochemical revolution has taken place in other fields, bread making (which has to be the oldest biochemical process known to man) has been converted from a biochemical process into a chemical and mechanical process.

The Chorleywood process suits large plant bakeries as it is capital intensive. It can use lower quality flour than more traditional methods and gives a higher yield of bread from a given weight of flour. The higher yield of bread occurs because the bread has a higher water content and less of the flour is used in feeding the yeast. Figure 6 shows a white loaf made by the Chorleywood process (a close up is given in Figure 7).

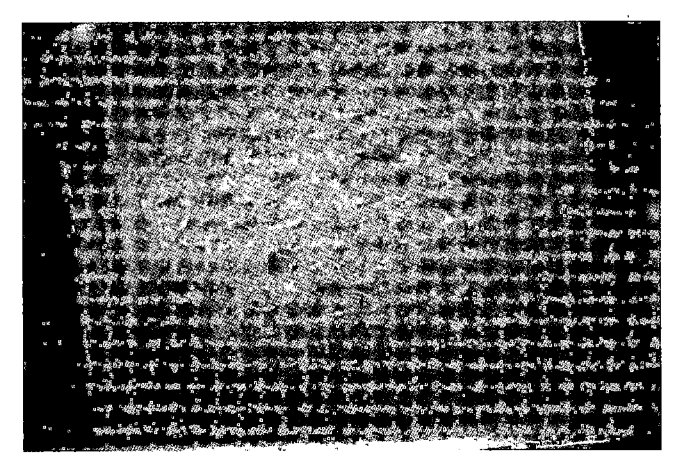

Figure 6 *A white loaf produced by the Chorleywood bread process (note the very even bubble size)*

Figure 7 *Close up view of a loaf produced by the Chorleywood bread process*

The Chorleywood process is chosen by bakeries that decide to make the cheapest possible bread. This sort of product has not enhanced the reputation of the product.

While the Chorleywood process can function with a lower protein flour than other processes it does require a flour with a minimum Hagberg Falling Number of 200, particularly where sliced bread is being made. If flour with too low a Hagberg Falling Number is used the dough will probably process but the finished loaves may jam in the slicing machine. It is not unknown for these loaf slicing machines to be jammed with a sticky mess of degraded starch that can only be removed with a pickaxe.

One of the advantages claimed for the Chorleywood process is that it allows bread to be made from flour produced only from British wheat. This claim holds true in most years, except when there is a particularly bad harvest, *e.g.* as in 1987. The effect of the CBP and the CAP on Canadian flour tonnages are as follows: imports were 2.5 million tonnes in the 1960s and are 300 000 tonnes today.

The Chorleywood process has spread from the UK to Australia, New Zealand and South Africa. It is relatively little used in the USA for several reasons. The cost advantages of using lower grade flour are less in the USA than in the UK while the higher ambient temperatures in the USA mean that a Chorleywood plant would have to be heavily cooled.

7.2.5.1 Method. The method below is given for a large plant bakery making loaves of sliced bread.

Mixing and Dough Development. In the Chorleywood process this is one step and only takes 2–5 min. The original specification for the process required an energy input of 11 W h per kg of dough. This requires a very specialised mixer. It has emerged subsequently that the optimum energy input depends on the particular flour. The dough must contain a higher level of ascorbic acid, yeast and water than a bulk fermentation dough would, as well as a hard fat.

In a large plant bakery the ingredients would be fed to the mixer in batches of around 300 kg. The mixer temperature is controlled at 28°C.

Dividing. When mixing is complete in 2–5 min the mass of dough is tipped into a divider. This divides the dough into individual pieces that are made into a ball shape. The weight of the pieces has to be tightly controlled to comply with weight legislation without giving away bread.

First Proving. The dough pieces then prove in a conveyor prover for 8 min.

7.2.6 Activated Dough Development (ADD)

This method offered the small baker the opportunity to produce no time bread without buying a specialist mixer merely by using an improver containing L-cysteine hydrochloride. ADD not only avoids the use of a specialised mixer, it will work with any mixer, even the sort of mixer that bakers normally keep for making cakes. The flavour of the resulting no time bread can be improved by adding some sour dough to the mix. Thus a small baker can make a first batch of bread by bulk fermentation and make later batches by ADD, adding sour dough to restore the flavour if required.

ADD methods needed a baker's grade flour of 12% protein, the sort of flour that would be used in a bulk fermentation process rather than that a CBP process would use. This is exactly the sort of flour that a small baker would use for a bulk fermentation process. Thus both processes could be used with the same grade of flour.

The ADD method was used by in-store bakeries and hot bread shops as well as small bakers. At one time approximately 5–10% of all commercially made bread could have been made by the ADD method.

The ADD method was founded on the use of the relatively rapid acting reducing agent L-cysteine in combination with the relatively slow-acting oxidising agent potassium bromate. The L-cysteine works by allowing the uncoiling and re-orientation of wheat proteins while the slower acting oxidising agent potassium bromate caused the formation of cross-links that allowed the three-dimensional structure of the gluten network to form.

Apart from flour the recipe would be: 2% yeast and 0.7–1.0% fat. The same amount of extra water is added as in the CBP process.

The flour improver would contain sufficient L-cysteine hydrochloride to give 35 mg kg^{-1} of flour (equivalent to around 27 mg kg^{-1} of L-cysteine) with sufficient potassium bromate to give 25 mg kg^{-1} of flour and sufficient ascorbic acid to give 50 mg kg^{-1} of flour. The above assumes a flour of 12% protein that has had added to it up to 20 mg kg^{-1} potassium bromate. Alternatively, with an untreated flour all the potassium bromate would be in the improver.

When potassium bromate was struck off the UK permitted list in 1990 the ADD process was no longer viable. Unsuccessful attempts were made to use ascorbic acid with or without azodicarbonamide. ADD then is only a viable method where potassium bromate is allowed. As the continued use of potassium bromate comes under further pressure then scope for using the ADD method decreases. Those bakers who had used ADD almost certainly did not go to the bulk fermentation process but moved to using a spiral mixer.

7.2.6.1 Method: Mixing and Dough Development. The ingredients, including an improver containing L-cysteine hydrochloride, are mixed and the dough develops in the mixer. While development is less rapid than in the Chorleywood process it is much faster than a bulk fermentation. Typically, mixing would take 10–20 min, with a final dough temperature of 28–30°C.

7.2.7 The Spiral Mixer Process

A further development from ADD methods is the use of spiral mixers by small bakers. These machines put energy into the dough less rapidly than the sort of mixer used in the Chorleywood process but a spiral mixer is less expensive and more versatile. The actual dough development time will always depend on the recipe, the flour improver and the flour but, for comparison, while a Chorleywood mixer will develop a dough in 2–5 min a spiral mixer will take 8–15 min.

The spiral mixer process is sometimes called the continental no time process because these mixers originated in continental Europe where small bakers are much more common. Neither L-cysteine nor potassium bromate are needed. The only flour treatment needed is ascorbic acid.

7.2.7.1 Method

Mixing. The ingredients are mixed in the bowl of the mixer using the mixer's low speed. The dough is then ready for development.

Dough Development. The dough is then developed by running the mixer to its high speed, typically for between 8 and 14 min.

7.2.8 Other Mechanical Dough Development Methods

There are two other much older methods of making bread that rely on mechanical dough development. One is exhibition bread the other is West Indian bread.

Exhibition bread is a product of no commercial significance whatsoever – it is the sort of bread made by apprentice bakers for bakery exhibitions at trade shows. The purpose of the method is to produce a completely uniform product to be judged by an arcane set of criteria. These criteria are of no importance to the bread buying public.

West Indian bread making, in contrast, was presumably developed to cope with high ambient temperatures when mechanical refrigeration was not available. Attempts to use the sort of bulk fermentation used in the UK would cause problems because of the high dough temperature.

7.2.8.1 Exhibition Bread

Mixing. The ingredients are hand mixed to make a dough.

Dough Development. The dough is rolled out by hand, then rolled up and re-rolled until it is felt to have developed.

7.2.8.2 West Indian Bread

Mixing. The dough is mixed conventionally. A very low dosage of yeast is used compared with conventional methods.

Dough Development. The dough is developed mechanically by rolling and re-rolling. This is often done by using a pastry brake but must originally have been done by hand.

7.2.9 Continuous Processes

Unlike other industries, bread making is moving away from continuous bread making processes and towards batch processing. Continuous processing never has the advantages in the food industry that it has in the chemical industry.

In the chemical industry processes are often capital intensive with consistent raw materials being converted into long-life products. If a process is capital intensive it pays to run the process 24 hours a day, 7 days a week, 52 weeks a year, as far as technically possible. Labour costs are likely to be minimised by working continuously as well.

In the food industry there is a need to stop and clean the plant for hygiene reasons. Food raw materials do vary and the need to take corrective action in terms of the process or the recipe often arises.

Bread is a short-life product that is neither sold nor consumed continuously. The need with bread is to get the product to the retail outlet in time to sell it.

Continuous dough-making processes do exist. One example is the Wallace and Tiernan Do-Maker process. This is not a continuous process in the sense that flour and other ingredients are mixed together and bread emerges at the far end, but it is a way of making dough continuously by mixing flour and a liquid pre-ferment or "brew". The pre-ferment is made by mixing a sugar solution with yeast, salt, melted fat and oxidising agents. This pre-ferment would be fermented for 2–4 hours, which obviously is not a continuous process.

Measured quantities of the pre-ferment are mixed with flour. The dough is then subjected to the combination of intense mixing and the action of the oxidising agent. It is then extruded and cut into loaf sized portions, proved and baked. Bread made by the Do-Maker process has a very even crumb texture, which is characteristic of the process.

The one stage of the baking process that can easily be worked continuously is baking. The bread is fed into a tunnel oven and emerges baked at the other end.

The AMFLOW process is similar to the Do-Maker process but the mixing chamber is horizontal rather than vertical. The pre-ferment stage is more complicated as it is multi-stage.

The Do-Maker process is believed to have been first used in the UK in 1956, before the CBP was invented. In 1969 around 35% of American bread was made by continuous processes. This percentage has been declining since the 1970s. Neither the Do-Maker nor the AMFLOW process is currently used in the UK.

The sort of British bakeries that might once have used the Do-Maker or AMFLOW process have almost certainly changed to the CBP. American bakeries have almost certainly not changed to the CBP process. The CBP process is little used in the USA and mostly not in mainstream bakeries. One reason is that American bakers claim that the bread does not taste the same. Another problem is that average ambient temperatures are higher in the USA than in the UK, meaning that American bakers would be likely to need to cool the mixing stage. This would be an unpopular extra expense and complication, not least because the cooling system might fail.

One of the big advantages of the CBP to the British baker, that it can use lower quality flour, is worthless in the USA as high quality flour attracts less of a price premium there. The ability to use flour milled from all English wheat was a particular advantage when the levies on non-EU wheat were at their highest.

7.2.10 Emergency No Time Process

This is very much a method of last resort that would be used when there has been some major failure in the bakery its machinery, staff or suppliers. The management would be faced with the stark choice of making bad bread or no bread at all. This is done in the hope that if the customers buy bad bread on one day they might not go elsewhere for their bread. If the customers have no bread they will have to go elsewhere and might not come back.

The method uses extra yeast and a higher temperature with just a single proof before baking. The finished bread is not very pleasant as the crumb structure is coarse with thick walls, and also it stales rapidly.

7.2.11 Gas Injection Processes

The idea of replacing the action of yeast with water aerated with carbon dioxide gas has been around for a long time. Dr Dauglish of Malvern produced such a process in 1860 that was claimed to avoid manual kneading. The resulting Aerated Bread Company had a plant designed by Killingworth William Hedges.

This sort of process works because the solubility of carbon dioxide in liquids increases with increasing pressure. This is true for all gases but carbon dioxide is much more soluble in liquids than the permanent gases because it is nearer its critical point.

If this type of system is to be used to make bread either the dough has to be saturated under pressure with carbon dioxide or a dough has to be made under pressure from water previously saturated with carbon dioxide. In either case when the dough was released to atmospheric pressure the water would be supersaturated with carbon dioxide which would cause dough to foam up. In Dauglish's process the dough was divided immediately after expansion and baked. The entire process, including baking time, took 90 min.

The Oakes Special Bread Process is a more modern continuous process in which carbon dioxide is injected into the dough. Neither of these processes is, as far as is known, in current use to make ordinary bread. Any process that raises the dough by injecting gas without yeast will, if there is no fermentation, produce a tasteless product. Significantly, Dauglish found the need to incorporate some fermented material in his product.

7.2.12 Part-baked Loaves

This is bread that has been baked to the point where the crumb has set and oven spring has occurred but before the Maillard reaction causes the crust to brown. The bread is then packed.

The idea was to offer the customer a way of having fresh baked bread early in the morning without it having to be delivered or fetched.

This product has probably lost a lot of its market to the domestic bread machine, which offers fresh baked bulk fermented bread. The retail market for these product seems to have shrunk to speciality products such as rolls or baguettes.

There is a trade in frozen part-baked breads. Once again it centres around speciality and difficult to make products such as baguettes. The frozen dough is used by hotels and some in-store bakeries. The excess cost of the process can be supported on these products in these circumstances.

7.2.13 French Bread

The long thin loaves of French bread are regarded in the rest of the world as an icon of France, alongside the Eiffel Tower. Their origin is said to be that one of the Austrian queens of France demanded the sort of loaf that she was accustomed to in Vienna. Possible candidates for the queen would be Anne of Austria, wife of Louis XIII, or Marie Antoinette, wife of Louis XVI. As central European wheat is hard, resembling North American wheat, this is a formidable problem with only French soft wheat available. A modern bakery technologist would find this difficult.

A Vienna is a long thin loaf, but of course made from central European hard wheat flour. Working with the available raw materials the French king's bakers succeeded at least in producing long thin loaves.

The long thin loaves we currently recognise as French bread are supposedly descended from this. In France these loaves are known as pain de Paris (Paris bread) as they were first available in Paris. In country districts the pain de campagne (country bread) was available. Pain de campagne is still made and is usually a large batch loaf. It can be a ring known as a Couronne (crown).

The products sold in the UK as "French sticks" or baguettes have become far more authentic. At one time the product sold as a French stick in Britain could be nearer to a Vienna than a French stick. Now most of these loaves are labelled as baguettes, although some of them are correctly sold as ficelles. In France, there is a whole range of long thin loaves, some are baguettes others are ficelles and yet others are batards.

All of these loaves will have been made from French soft wheat flour without the use of fat or soy flour. This flour will have been milled with a low starch damage from varieties of French soft wheat grown for bead making.

The method used to make it will vary. The small craft bakery is much more common in France than in Britain, and small bakers tend to have their own way of doing things. The usual method would be a sponge batter method with some added sour dough. Some bakers do not add the sour dough but use a long fermented batter, *e.g.* fermented overnight in the sponge batter process.

A substantial proportion of this sort of bread made in France is now made in plant bakeries. The same sort of social forces are pushing French people to one-stop shopping as are operating in the UK.

The demand for authentic French bread is obviously a sign of greater cosmopolitanism fuelled by increased foreign travel. The commonest British attempts at an authentic French loaf are made from flour milled in Britain using French wheat. The requirement to mill with a very low starch damage is not achievable in some British bread flour mills.

Typically this sort of product would be made using a spiral mixer with a specially formulated bread improver. The other requirements are the moulds to shape the dough and an oven with steam.

7.3 OTHER BREADS

7.3.1 Brown and Wholemeal

Brown and wholemeal bread are generally made in a similar way to white bread except that a higher level of fat is normally used, *e.g.* 1.5% of the flour weight as fat, compared with 1% for white bread. In making wholemeal or brown bread by the CBP process the fat level must be raised. It is only possible to make wholemeal bread by the CBP because

Figure 8 *A wholemeal loaf produced by the Chorleywood bread process (note the very even bubble size)*

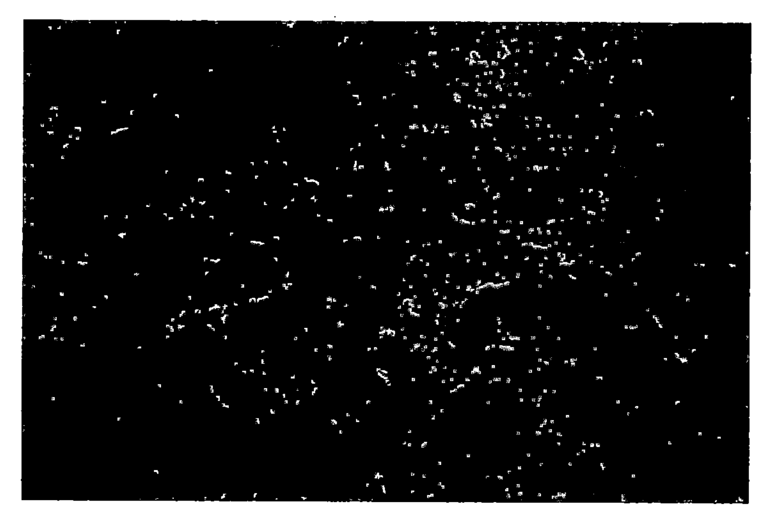

Figure 9 *Close up view of a wholemeal loaf produced by the Chorleywood bread process (note the very small, even bubble size)*

the regulations were modified to allow the use of ascorbic acid. Figures 8 and 9 show pictures of CBP-produced wholemeal bread.

In general wholemeal bread is made with shorter fermentation times than white bread. As wholemeal flour has a higher water absorption than white a higher water addition can be used.

Although a miller will always get a higher yield of wholemeal flour this is offset because in practice wholemeal flour needs a stronger grist than white flour. Since the protein content of the wheat berry is higher towards the outside a given grist will always give a higher protein wholemeal flour than when milled to produce a white flour.

Wholemeal and brown flours create problems for both millers and bakers since both sorts of flour have a limited shelf life (about three months) compared with the twelve month shelf life of white flour. Steps have to be taken to prevent brown or wholemeal flour contaminating white flour.

Demand for wholemeal bread has increased in recent years, presumably on the basis of the health benefits. However it is made, wholemeal bread is unlikely to expand as much as white bread since the bran particles tend to burst the gas bubbles. The wholemeal bread that has the greatest expansion is that produced by the CBP process. This product is the nearest to a wholemeal loaf with the texture of white bread.

A larger proportion of wholemeal than white bread is still made by bulk fermentation methods. One reason for this is that although old fashioned wholemeal bread is a minority taste the minority who consume it like it that way. They like the close texture and flavour of the traditional product.

7.3.2 Wheat Germ Breads

These are normally sold under a trade name. One such trade name is heavily advertised and the brand owner has started using it for its white bread.

The flour is made by adding at least 10% of heat-treated wheat germ to a white flour. The purpose of the heat treatment is to stabilise the wheat germ oil.

7.3.3 High Protein Breads

These products are made by adding protein to flour. While whey protein, soya protein, casein and yeast can be used the protein normally employed is pure vital wheat gluten.

Commonly the products are sold to slimmers under a trade name as carbohydrate or starch reduced. Of course the product contains less carbohydrate, *i.e.* starch than ordinary bread because the percentage of protein has been increased.

7.3.4 High Fibre and Multigrain Breads

These are breads that have had various grains and fibres added, usually with some nutritional benefit alleged.

7.3.5 Soft Grain Breads

These products were developed to offer the consumer a fibre-enhanced bread that still tasted like white bread rather than brown or wholemeal. Various kibbled grains, *e.g.* kibbled rye and kibbled wheat are added. The addition can be made either in the bakery or at the flour mill.

If the addition is made at the flour mill this grade of flour can be supplied directly to the bakery for use. This solution suits the large plant bakery and the domestic baker. Adding a concentrate to a mix derived from a bag of baker's flour is a solution that suits a small baker or a supermarket in-store bakery. There is then no need to stock an extra grade of flour for a product that will probably have a small sale.

7.3.6 Ethnic Multigrain Breads

These products use other grains in the recipe. Some of these products are made in the UK for people of the ethnic origin concerned while others are made as made as a way of producing a different tasting high fibre loaf. The authenticity of the latter type might not be great.

7.3.7 Slimming and Health High Fibre Breads

These products add a range of ingredients to bread with a view to obtaining some health benefit. The simplest type to understand are those that add cellulose. Cellulose can not be metabolised by humans so these ingredients add bulk to bread while reducing the amount of available energy.

The idea is that the slimmer eats the same quantity of bread as before but the bread's reduced energy content helps to reduce the slimmer's energy intake. The reduced energy intake should then cause the slimmer to lose weight.

The materials used, subject to the legislation, would be a delignified α-cellulose probably derived from wood or sodium carboxylmethyl cellulose. The level of usage would be of the order of 5–10%. At the time of writing, cellulose-containing slimming breads are legal in both the USA and the UK. A typical UK labelling requirement would be "Can only assist weight loss as part of a calorie controlled diet". This statement is intended to deter those who believe that there is a magic way to lose weight.

The other class of health high fibre breads have ingredients such as soya and linseed. The high soya content of these products has been promoted as making them especially suitable for women, particularly middle aged women because of the high content of oestrogenic hormone mimics. Curiously, they are not labelled as being unsuitable for men.

Linseed is one of the few vegetable sources of the omega 3 class of essential fatty acids, hence it is ideally suited for this type of product. The combination of linseed and soya produces a bread that has a low glycemic index, which is another bonus for this sort of product. This reduced glycemic index makes the product more attractive to slimmers and diabetics.

7.3.8 Bread with Added Malt Grains

This type of product is widely available as a slightly different taste and texture added value product. It is often available under a trade mark, which the trade mark owners rigorously protect, even by sending their staff out to ask for the product and taking action if a competitor's product is supplied instead. The trade mark owner is trying to prevent their trade mark being devalued to merely the generic term for the product.

The miller supplies the flour to bakeries and at retail on the basis that the baker can then sell the bread under the trade name. Similar bread not made from the trade marked flour would have to be sold either under another trade mark or under a generic description such as "Bread with added malt grains".

The technical issues with these products are as follows: The malt must not introduce sufficient starch splitting enzymes to make the product unhandleable. The base flour used has, therefore, to be a strong flour with a low level of amylase activity. This requirement makes the product expensive.

The taste of the product includes "nutty" flavours that have been produced by Maillard reactions between reducing sugars produced by the malt and proteins present. These flavours and the included grains give the product its distinctive character.

7.3.9 Bread Containing Cereals Other than Wheat

While some of the products considered above do contain cereal other than wheat it is a small proportion of the whole. In this section the rye breads are either wheat free or contain only a small proportion of wheat. These breads are made either for an agricultural reason, *i.e.* making bread where wheat will not grow, or to avoid wheat for a health reason. Bread made with a proportion of wheat is known as composite bread.

7.3.9.1 Rye Breads. The British consumer, in general, does not like the taste of rye bread. Most of the rye bread made in the UK is for the benefit of immigrant communities that traditionally eat rye bread. In Scandinavia, Central and Eastern Europe, rye breads are eaten as a delicacy not as a substitute for wheat bread. It has been recorded that the Vikings brought rye bread to Britain but they were an immigrant group.

Types of Rye Bread. The usual descriptions are logical. Rye bread is bread made solely from rye flour, rye/wheat bread contains a minimum of 50% of rye flour while wheat/rye bread contains not less than 50% wheat flour with not less than 10% of rye flour. Some recipes for San Francisco sour dough bread have a proportion of rye flour. No doubt, originally, the prospectors used what ever was available.

Rye and Climate. Rye will grow in conditions where wheat will not grow, so rye is available where wheat is not. Fortunately, rye is the only temperate cereal whose proteins can develop in a similar way to wheat gluten.

Where rye is grown the climate can be such that the rye grains germinate in the field. The phenomenon is not unknown with wheat but is more likely in more marginal climates.

The effect of using germinated grain is that the enzymic level of the flour rises. Rye flour with a Falling Number below 80 gives loaves with a sticky crumb and is best avoided. If the Falling Number is in the range 90–110 the rye flour can be used with additives such as an acid to adjust the pH to 4.0–4.2, increasing the salt to 2% on flour weight, using an emulsifier at 0.2–0.5% of flour weight and adding 1–3% of the rye flour weight of pre-gelatinised wheat flour.

Rye Proteins. While rye is the only European cereal able to completely replace wheat in bread, rye protein is not as effective as wheat protein. One reason for this is that as much as 80% of the protein in a rye sour dough is soluble compared with 10% of soluble protein in a wheat dough. One factor that inhibits the formation of a gluten-like complex is the 4–7% of pentosans present, which bind water and raise the viscosity of the dough. The crumb structure is then formed from the pentosans in combination with the starch.

Rye Starch. Rye starch gelatinises at or around the temperature at which α-amylase has its maximum activity (55–70°C). α-Amylase activity tends to be high in rye flour so steps have to be taken to minimise it. One step is to acidify the dough either by adding acid or using a sour dough process where the acids are produced by fermentation. The other measure is to use 2% of salt on the flour weight.

7.3.9.2 Straight Dough Process for Rye Breads. Here the dough is acidified with lactic acid or by adding acidic citrates. Yeast is added as in conventional bread making. It is necessary to mix the dough slowly lest the high viscosity of the pentosans should toughen the dough.

7.3.9.3 The Sour Dough Process for Rye Bread. As with wheat-based bread, sour dough was the original way of making bread before manufactured yeast became available. The production of acids during the sour dough process is convenient since rye bread needs to be acidified to encourage the pentosans to swell as well as inactivating the amylases. Part of the flavour of sour dough bread comes from the formation of acetic and lactic acids produced in the process.

While it is always possible to make a sour dough by leaving some flour and water to acquire suitable micro-organisms a working sour dough bakery would keep its own culture going by holding back a portion of the culture.

The alternative to the above is to purchase a sour dough starter culture. A whole range of starter cultures is available, varying from those that merely claim to produce a sour dough to those that claim specific origins. Examples of origin cultures are San Francisco, Klondyke, Middle East, Paris, German, Austrian and Russian.

7.3.9.4 Starting a Culture. A culture would be started by leaving a rye dough to stand at 24–27°C for several hours, which is likely to induce the grain microorganisms to start a lactic acid fermentation. An alternative is to add sour milk to the dough followed by resting the dough for a few hours. A mixture of pure organic acids can be added to simulate the flavour of a proper sour dough.

If the culture is to provide both the yeast and the flavour of sour dough then either it must acquire a wild yeast or a starter culture that includes yeast must be added. In some cases the sour dough culture is only used to give the sour dough taste while conventional yeast is added.

If a started culture is used the culture is activated by mixing it with rye flour and water and leaving it to stand in a warm place until the culture is fully active. The active culture is then kept going by feeding it flour and water.

When the culture is fully active the culture is mixed in with flour, water, salt and any fat. The resulting dough is kneaded carefully to avoid too much toughening. The dough is then fermented say for half to one hour, knocked back, scaled, proved and baked.

Some sour dough bread is made by using commercial yeast but with a proportion of genuine sour dough. Ordinary baker's yeast is at a disadvantage in rye sour dough because the low pH that is essential for rye bread is not the optimum pH for the yeast.

Conventional improvers are not used in rye bread but additives are sometimes used to increase the water absorption of the dough. Examples are polysaccharide gums such as guar and locust bean gum as well as pregelatinised potato flour, rice starch or maize starch.

7.3.9.5 Pumpernickel. Pumpernickel is a special type of rye bread. It is black with a soft chewy texture and a pungent flavour. It is made from a very coarse rye flour by a sour dough method. The baking time is very extended, starting at 150°C and finishing around 110°C. Because of its long shelf life pumpernickel is normally packaged in aluminium foil.

7.3.10 Crispbread

7.3.10.1 Rye Crispbread. There are two forms of rye crispbread. White crispbread is unfermented while brown crispbread is fermented with yeast.

Process for Unfermented Rye Crispbread. The original Swedish way of making this product was to mix rye flour or rye meal with snow or powdered ice. The product is then aerated by the expansion of the air bubbles when the icy foam is placed in the oven.

Process for Fermented Rye Crispbread. This process is a bulk fermentation process with a dough made from wholemeal rye flour, water, yeast and salt. This dough is fermented for 2–3 hours at 24–27°C. Then the dough is "knocked back" by mixing for 5–6 min followed by proving for 30 min.

After this the dough is rolled into sheets, dusted with rye flour and cut into pieces. The pieces are then baked for 10–12 min at 216–249°C. The product is finally dried to below 1% moisture by standing the baked piece on edge in a drying tunnel for 2–3 hours at 93–104°C.

7.3.10.2 Extruded Crispbread. These products are of recent origin and were originally made from wheat flour. The original product was sold under a trade name. The product is made by mixing white flour and water in an extruder and then applying intense shear that heats the mixture. As the pressure in the extruder is above atmospheric the boiling point is raised above 100°C. The gelatinised starch is then extruded through a slit. Once the pressure falls to atmospheric the water present flashes off to steam, rapidly cooling the product, which sets and can be cut into pieces.

7.3.10.3 Triticale. This grain is a cross between rye and wheat. The obvious aim is to produce a grain having the bread making properties of wheat and the hardiness of rye. The problem is not to produce varieties with the bread making properties of rye and the hardiness of wheat.

Early versions were not very successful but later versions were more satisfactory. Bread baked from a mixture of 65% wheat flour and 35% of stone ground wholemeal triticale has been marketed in the USA.

Triticale flour has been extensively tested in Poland, a country where rye bread is traditional. The best results were obtained by using 90% triticale flour with 10% rye flour. The rye flour was made into a flour brew for 24 hours at 28–29°C. Half the triticale flour was made into a sour dough for 3 hours at 32°C followed by mixing with the rest of the ingredients plus 1.5% of salt on the flour weight. The bread was then scaled and proved for 30 min at 32°C followed by baking at 235–245°C.

At the time of writing, most of the triticale that is grown is grown for animal food.

7.3.10.4 Barley and Oats. While barley and oats do not have any bread making properties they can be added to bread. Obviously introducing amyltic activity with the barley is to be avoided. Similarly, oats can introduce undesirable enzymic activity. Originally, oats and barley found their way into bread because of shortages; now oats are likely to be incorporated into bread because they are believed to be healthy.

During World War II the British government allowed up to 10% of the grist for the national loaf to be barley or oats and barley. This maximum substitution occurred in 1943, when supplies of imported food had fallen to two months. At that time of course the UK was dependent on imported wheat for bread making.

The husk of the barley was removed and the oats were used as dehusked groats.

Bread has been made in Norway from wheat flour 78% extraction (50 parts); barley flour 60% extraction rate (20 parts); wholemeal flour (30 parts); with extra fat added.

In the USA, specially treated oat bakery ingredients are produced by heating the grain to 100°C with steam and leaving the grain in silos for 12 hours. This process conditions the oats so that they can then be dehulled.

7.3.10.5 Rice. Rice can be made into bread either in combination with wheat flour or in its own. A mixture of wheat and rice flour might be used where rice is available but wheat will not grow to minimise imports of wheat.

A product resembling bread can be made using 75% of wheat flour and 25% of flour milled from extruded rice. The importance of extruding the rice is that some at least of the starch would have gelatinised.

Bread made from rice but free of wheat has a low content of sodium, protein, fat and fibre. This makes rice bread suitable for those who suffer from inflamed kidneys, hypertension and coeliac disease subject to medical advice.

In the absence of gluten some other system must be used to cause the bread to rise. One possibility is to use hydroxypropylmethyl cellulose, which forms a film with the rice flour and water that traps the gases and acts as a substitute for gluten.

Another use for rice in bread is rice bran. Rice bran is a by-product of producing white rice. It is claimed up to 15% of rice bran can be added to baked goods, increasing water absorption and adding amino acids, vitamins and minerals. All of this is achieved without affecting loaf volume, mixing tolerance or dough fermentation. It is also claimed that rice bran lowers serum cholesterol but use of this claim would be subject to legislation.

7.3.10.6 Maize. Bread is made from maize flour in Latin America. Mexican tortillas are an unleavened maize bread, *i.e.* they are essentially chapatis made from maize rather than wheat flour.

Maize proteins do have the ability to develop in a similar way to wheat. The use of maize in Latin America is not too surprising as the plant is native of that region.

7.3.10.7 Composite Flour. Composite flour is flour made by mixing wheat flour with flour milled from some other grain. The usual reason for doing this is in a developing country where wheat will not grow so as to minimise wheat imports. The wisdom of encouraging bread consumption in a country where wheat will not grow and there is no tradition of consuming wheat-based bread has been questioned.

Typically, the wheat flour would be blended with maize, sorghum, millet, rice or cassava.

All of these cereals and cassava, which is a root crop, will grow where wheat will not. Sorghum will grow in particularly unfavourable conditions.

The non-wheat ingredient does not improve the bread but merely makes the wheat flour go further. Typical substitution levels are sorghum flour and millet flour 15–20% of the wheat flour; maize 20–25% of the wheat flour.

In Senegal and Sudan, bread is being made from a blend of 72% extraction wheat flour (70%) and 72–75% extraction white sorghum flour (30%).

When making these products extra amounts of water often have to be added to compensate for the increased level of water absorption of the wheat substitute.

7.3.10.8 Distillers Spent Grains. Distillers spent grain is the solid residue left after the grain has been fermented to make spirits, typically whisky. This residue is the unfermentable parts of the grain and is high in fibre. Traditionally, the only outlet for this material was as cattle food or as an effluent. In addition to the fibre, protein, fat and the insoluble vitamins and minerals are present.

7.3.11 Bread for Special Dietary Needs

The main aim of these products is to produce a bread that can be eaten by those suffering from an allergy to gluten. This disease is known as coeliac disease. An estimated 1% of the British population suffer from coeliac disease, with only around one fifth being diagnosed.

There are also those who believe that they have an intolerance to gluten, *i.e.* a problem with digesting gluten, but without the full immune response to gluten found in coeliac disease. Coeliac disease is found in all age groups but its occurrence is greatest in the elderly.

In either coeliac disease or gluten intolerance there is a need for products that do not contain gluten. In the case of special dietary needs the ordinary food laws are set aside. This is of course an entirely logical position to take.

In the case of bread for coeliacs all possibilities of contamination by wheat products, even at a trace level has to avoided. There is also the problem of getting the loaf to rise in the absence of gluten.

A solution that has been made to work is to use various polysaccharides. Examples are the bacterial gum xanthan gum, carob gum, guar gum and gum acacia.

The favoured solution seems to be xanthan gum with rice starch and potato starch. Other systems use sorghum flour.

Carob gum, guar gum and gum acacia have the advantage of being natural. As gum acacia gives a lower viscosity than the other gums more has to be used. However, as this gum is more soluble it is possible to use more.

Another class of gluten substitute is the modified celluloses dimethylglyoxime cellulose and hydroxyl propyl methyl cellulose; 1.6% of carboxymethyl cellulose or 3% of hydroxy propyl methyl cellulose is added to rice flour or a mixture of rice flour and potato starch. The cellulose percentages are based on the dry weight of the flour substitute.

The xanthan gum, rice starch and potato starch system is available as a mix to be made in domestic bread makers.

7.3.12 War and Famine Breads

In times of war and famine bread has to be made from what ever is available. Some of the most desperate bread making was during the siege if Leningrad in World War II. The defenders were reduced to a recipe with 50% rye flour with sawdust and cotton seed added.

7.4 OTHER VARIANTS OF BREAD

7.4.1 Flat Breads

These products are the sort of bread that can be made without an oven. If a flat sheet of dough is placed on a hot griddle or bake stone then the dough will cook by conduction. The floor or wall of a heated oven or a modern deck oven are other possibilities.

The traditional product would be based on sour dough. The day's bread would be made by adding flour and water to the sour dough starter. Before baking, a portion of the dough is kept as the next day's starter. This procedure is the only one possible if manufactured yeast is not available. Sour dough processes also have an advantage in hot countries because the acidity discourages spoilage organisms.

7.4.1.1 Origins. These products are common throughout the middle East and the Indian sub-continent. An example is the Indian Naan bread. Naan bread has spread from British Indian restaurants to British mainstream supermarkets. The standard accompaniment to a balti curry is Naan bread (Figure 10).

The cuisine of the countries where flat breads are eaten consists of curries and purées, which can be eaten conveniently by scooping them up with pieces of flat bead. This is analogous to the mediaeval custom of serving the food on a trencher, a slice of bread that was eaten at the end of the meal.

7.4.1.2 Ingredients. A simple recipe consists of wholemeal flour, salt, water and sour dough starter. The flour would be stone ground, possibly by hand.

Alternatively, ground pulses, sesame seeds, fat (either ghee or sesame seed paste) and honey are used. The product could be flavoured by

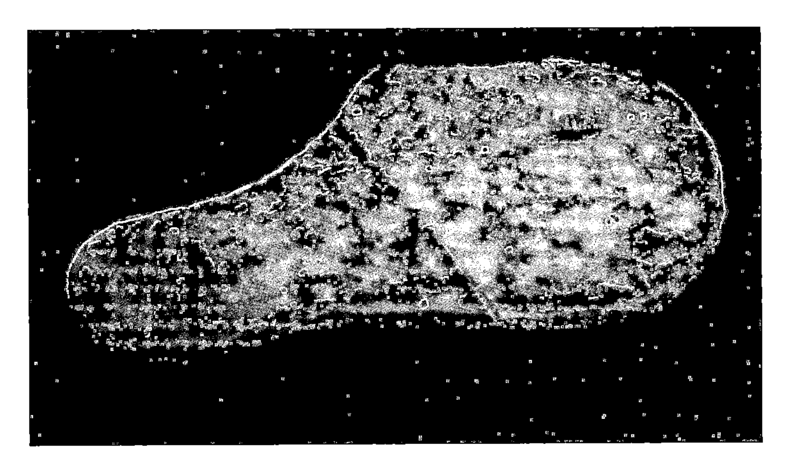

Figure 10 *A Naan bread*

adding anise, cardamom, curry powder or dill. While wheat is the major cereal present as much as 20% of the flour could be supplemented with ground rice, barley, maize or sorghum.

7.4.1.3 The Method. In domestic production the dough is mixed by hand. In a small commercial bakery the dough might be mixed in a low speed mixer and sheeted out using a mechanised sheeting roller, *i.e.* akin to a pastry brake.

The traditional oven in a bakery would be fired by building a wood fire inside the oven then raking the embers out and baking on the floor of the oven. A domestic oven or tandoor might consist of an upturned urn about one metre high with a half metre opening in the top and a small opening in the bottom to feed the fire. When the oven is hot enough the sheeted dough is inserted through the top and plastered on the walls. This operation is done by hand. With care, burning is avoided since, although the oven walls are around 250°C, dough is a poor conductor of heat. The baked bread is removed with wooden tongs. In the commercial hearth oven the bread is inserted and removed with a wooden peel of the sort once used by European bakers.

In the sort of British bakery that has been set up to produce Naan bread for supermarket and restaurant sales everything is automated. British consumers have been introduced to Naan breads in Indian restaurants so the supermarket product has to be of restaurant quality.

After the dough has been mixed it is divided into balls and proved. In a restaurant, Naan bread is made from pre-mixed dough so that the product is always well fermented. The manufactured product is proved for 20–40 min. The proved dough is then shaped and rested for 20 min.

The product is baked in a special clay-lined oven. Above the clay there is an arch of 6 mm of Inconel, a special alloy steel, to stand the heat. The oven is intended to give the same taste as a traditional tandoor. Unlike the traditional tandoor the oven is horizontal rather than vertical. There are five burners, three above the Naan and three below. The lower burners provide base heat as in a tandoor while the burners above throw heat across the surface.

The finished product is essentially sterile and if kept in a germ-free environment will be protected from spoilage. Some manufacturers handle the finished product in clean room conditions to enhance the shelf life.

The finished product is then packed. If the product is going to a restaurant it will probably be packed in boxes. Product for supermarkets is then flow packed with or without a carbon dioxide gas flush. Those types of Naan most susceptible to oxidation receive the gas flush. Suitably packed Naan will keep for three or four months.

7.4.2 Pitta Bread

This traditional Middle Eastern product has the unique distinction of being called bread yet is specifically excluded from the bread and flour regulations. The only consequences of this are that pitta (also spelt pita) bread does not have to be fortified and any food additive that is permitted in food generally but is not permitted in bread would be permitted in pitta bread. In practice, pitta bread makers tend to be rather conservative.

Typically wheat flour with a 75–82% extraction rate is used. The flour used in the UK is a strong bread flour.

Traditional pitta bread is made by a sour dough process but it can be made using manufactured yeast. If the sour dough process is used a relatively large portion of sour dough (approximately 20%) is used, giving a more rapid fermentation than most sour dough products.

The first proving step is about 1 hour. Next the dough is divided into pieces weighing about 75–100 g each. These are then re-proved and flattened into circles about 1 cm thick and 15–20 cm in diameter. These are then proved for 0.5–0.75 hour. The pieces are then subjected to a short time, high temperature cook of 1.5–2 min at 350–400°C. Under these conditions a crust is formed immediately and the pieces balloon as the air, water, ethanol and carbon dioxide expand. When the bread is removed from the oven it cools and collapses to the familiar shape but with a pocket inside. It is an interesting speculation if this product was originally made on a griddle. Figure 11 shows some pitta bread.

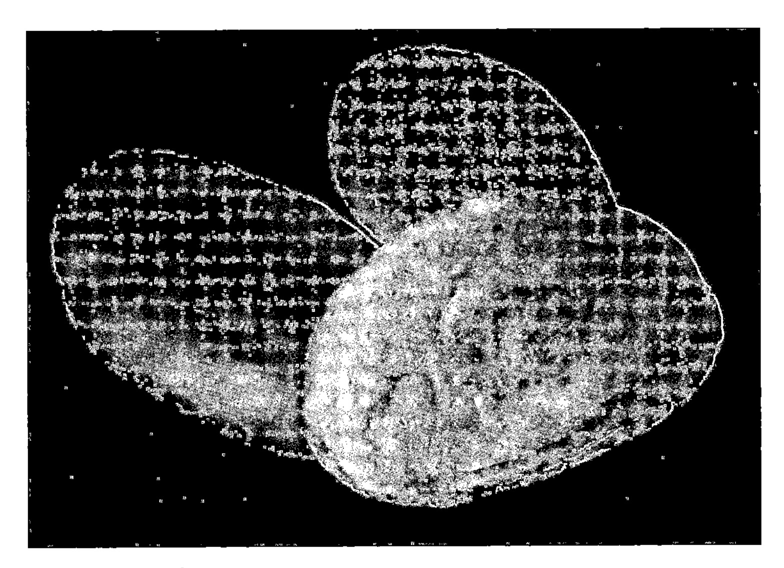

Figure 11 *Pitta bread*

7.4.3 Muffins

Muffins are the British contribution to the field of flat breads. This refers to the product known as an English muffin. American muffins are a chemically raised sweet product.

There is a view that muffins originated in Wales. They were originally a sour dough flat bread. They seem to have disappeared from British life, except in a nursery rhyme. Presumably the original product had a short shelf life. The sort of product now sold as an English muffin has overcome some of these problems. The modern product is a disc shaped product about 7–10 cm in diameter and 2 cm high.

A modern muffin plant uses a very soft dough baked in a covered griddle cup on a highly automated line. The dough must flow into the cup and lead to large holes when cooked.

A dough consists of 83–85% water on a flour basis, with 2% sugar, 1.5% salt, 5–8% yeast with 0.5–0.7% calcium propionate. The problem with this formulation is that it must be soft enough to flow into the cups but to prevent it sticking it must be handled cold around 20°C.

The dough is transferred from the mixer, divided, rounded and deposited from the intermediate prover into the griddle cups where the dough pieces undergo final proving. The product is then baked.

Most of the problems with muffins occur because the finished product has a high moisture content (around 45%) and has to be made from a high moisture dough. This dough gives handling problems and needs protection from mould infections. Handling problems are dealt with by using the dough cold and liberal applications of a dusting material such as a mixture of cornmeal and corn flour. An alternative dusting medium would be a specialised dusting flour.

The mould problem is tackled by using high levels of calcium propionate. This creates problems because it inhibits the yeast. A high level of yeast is used to try and counter this. An alternative way of inhibiting moulds is to spray the muffins with a potassium sorbate solution as they leave the oven. In this case the calcium sorbate is omitted from the dough and the level of yeast is reduced by 1%.

Alternatively, a sour dough base or vinegar can be added to the dough. Vinegar of course contains acetic acid, which apart from lowering the pH has antibacterial and mould-inhibiting properties. If vinegar and/or sour dough are used without calcium propionate then a "no preservatives" claim can be sustained. If vinegar is used with calcium propionate the anti-mould activity is enhanced by the low pH.

In general, muffins are expected to relatively tough and chewy with 3 to 6 mm holes, which are large compared with those in bread (Figure 12). The side wall should be straight and the corner with the flat top crust should be rounded rather than sharp (Figures 13 and 14).

The toughness is sometimes obtained by adding dried vital wheat gluten to the dough. This is one of the few products where this is done.

In some cases proteases are added to open up the crumb structure. This is more likely to happen in North America than the UK. The response in the UK would probably be to use a weaker flour.

Muffins can be made on a smaller scale; Table 2 gives a possible recipe.

7.4.3.1 Method. The ingredients are made into a dough that is allowed to ferment for 1 hour. The dough is then knocked back and fermented for a further 0.5 hour.

Next, the dough is scaled, moulded into shape, and the pieces are placed on a board that has been dusted with maize polenta or rice cones. The pieces are then proved for 0.5 hour. Next, the muffins are baked on a hot plate and flipped over to cook the other side.

7.4.4 Crumpets

While these products can be made by just depositing the batter in crumpet rings they are normally made on a continuous plant. One big difference between them and muffins is that crumpets are made from a batter rather than a dough.

A possible recipe is given in Table 3.

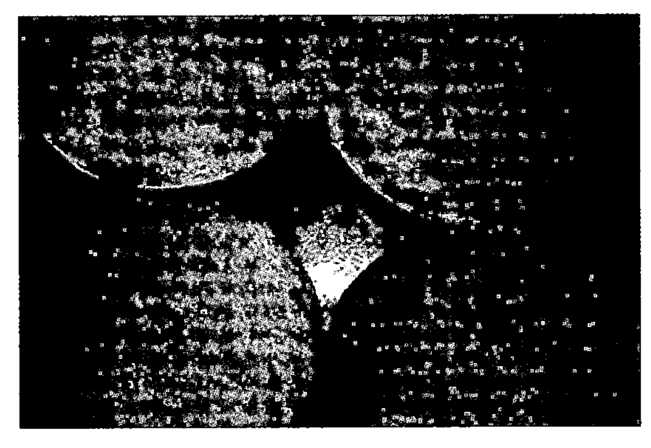

Figure 12 *Cross section of a muffin*

Figure 13 *Side wall of a muffin*

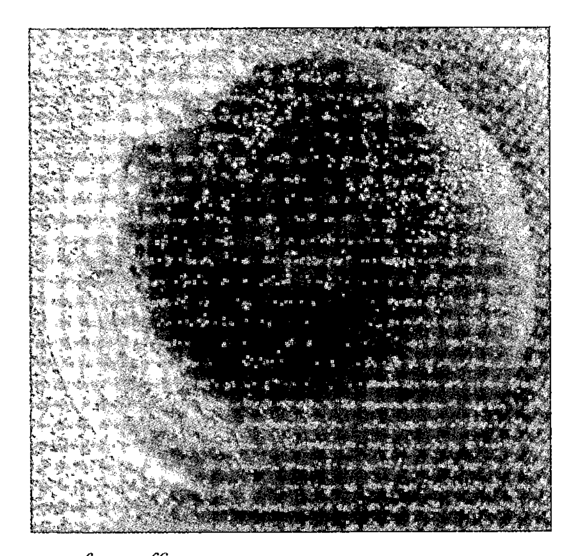

Figure 14 *Top crust of a muffin*

Table 2 *Recipe for muffins*

Parts by weight	*Ingredient*
1000	Strong flour
780	Water
25	Yeast
15	Salt
6	Sugar
10	Shortening

Table 3 *Recipe for crumpets*

Parts by weight	*Ingredient*
1000	Strong flour
30	Salt
7	Sodium bicarbonate
11	Yeast
1200	Water

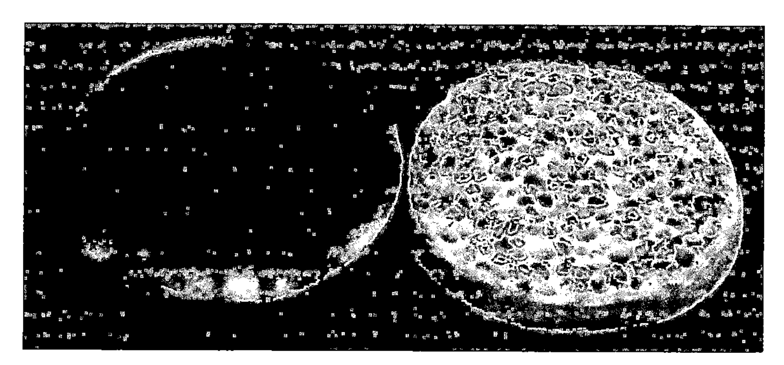

Figure 15 *Both sides of a crumpet*

7.4.4.1 Method. The ingredients are made into a batter that is fermented for 0.5 hour. In a plant bakery the batter is then deposited into hoops set on polished hot plates. A hot plate is then lowered down onto the surface to lightly bake the top of the crumpet. In hand making, this would be achieved by turning the crumpet over onto a hot plate. Figure 15 shows both sides of crumpets.

If the same sort of batter is deposited onto a hot plate with no hoops the product obtained is a pikelet. The pikelet is flipped over to brown the top. Pikelets are a delicacy in the Midlands. The name is said to be derived from the Welsh bara pyglyd or "pitchy bread".

Both crumpets and pikelets have high moisture content so mould growth can be a problem. Mould inhibitors or vinegar are sometimes used to counter the problem.

7.4.5 Pizza

There can be little doubt that pizza originated as a way of using up scraps of bread dough. Various origins are claimed for pizza, ranging to Naples in the nineteenth century, classical antiquity and the USA.

Pizza or something like it could have been produced by the Phoenicians, the Greeks or the Romans. Given the tendency of people to eat their food off a piece of flat bread the inventive step of cooking the other ingredients with the bread could have occurred to any number of people.

It is said that the soldiers of Darius the Great (521–486 BC) emperor of Persia baked a kind of flat bread on their shields and then covered it with cheese and dates when on campaign. Cato the Elder, *i.e.* Marcus Porcius Cato (234–149 BC), wrote of "a flat round of dough dressed with olive oil, herbs and honey baked on stones". Shops were found in the ruins of Pompeii apparently equipped for the manufacture and sale of flat breads, possibly pizzas.

The modern era of pizzas is normally taken as starting in 1889 with a visit by King Umberto I (1844–1900) and his Queen Margherita di Savoia (1851–1926) to their summer palace near Naples. They sent for one Raffaele Esposito, a leading pizza maker who produced a pizza in the colours of the Italian national flag, red white and green. The red was produced by tomato purée, the white was mozzarella cheese and the green was basil. This pizza was christened pizza Margherita in honour of the occasion.

Italian immigrants took the recipe to the USA, where made with American flour a different product emerged. The first American pizza parlour opened in 1905 in New York City. The next American development was the Chicago deep dish pizza in 1943.

Pizza seems to have arrived in the UK from the USA rather than via Italy – visiting American service men possibly had something to do with it. The effect of increased European travel and visits to Italian restaurants has led to a demand for authentic Italian pizza. The problem is, what is meant by pizza?

The classic Neapolitan pizza is supposed to be similar to Margherita but with added anchovies. This is a thin light product that is cooked by placing it on a very hot bake stone (around 600°C) on top of a wood fired oven for 0.5–1 min.

Pizza is, because of its history, one of the more variable bakery products. Pizzas made from Italian soft flour are never going to be the same as an American pizza made from American flour.

Chicago style pizza is made in a pan or dish with the cheese going in first and then sauce on top. The crust is then formed up the side of the pan – even with two crusts with sauce in between, known as a "stuffed crust".

The St Louis style pizza is a thin crust pizza using local Provel cheese in place of Mozzarella. This product is crisp with a seasoning of oregano, other spices and a slightly sweet sauce. Hawaiian pizza uses pineapple and Canadian bacon, giving a rather sweeter product.

British pizzas probably started from a point where they would be unrecognisable to either Italians or Americans. As in other fields the demand for authenticity has led to the production of products that are much closer to the original.

The variations possible are limitless; however, pizza essentially is bread dough with other ingredients added. If the pizza is to be thin a dough is needed that spreads rather than lifts. If the pizza is to be thick then the dough needs to be nearer to British or American bread dough.

The British practice for thin pizzas is to use flour based on a high proportion of soft English non-bread making wheat. This gives a dough that spreads but does not rise. Such a flour is likely to have a low Falling Number, *i.e.* an appreciable amount of amylase activity, which again assists in the spreading of the dough. If a thick pizza is desired a stronger flour would be used. American practice is to either use flours with approximately 12% or 14% protein, depending on the product. Alternatively, an American pizza maker might stock just the 12% protein flour and add dried vital wheat gluten to those products where the higher protein content is felt to be needed.

The processing given to pizza doughs is very variable, ranging from large plant bakeries to small pizzerias (pizza makers) where the dough is kneaded by stretching it and folding where the customers can see it.

Some pizza doughs are mixed, kneaded, given a short proof, and sheeted out for use. Other doughs are mixed, divided and rounded, followed by 24 hours in a retarder at 5–8°C. This gives the effect of a very long fermentation step. The dough is then sheeted and the pizza is made up and baked.

Lastly, frozen pizzas are often not yeast raised at all but are chemically leavened. The usual leavening agent would be a double acting baking powder.

7.4.6 Rich Dough Products

There are various of these products made around the world. They are yeast leavened but based on a dough that is high in fat and sugar, possibly with egg added.

7.4.6.1 Croissants. While these are thought of as archetypally French their origin is in Vienna. When Vienna was besieged by the Turks the bakers, as ever, were working in the small hours of the morning. As a reward the bakers were allowed to make a product in the shape of the Turkish crescent. These products were called Kipfeln. Marie Antoinette introduced them to the French court, where they became known as croissants.

The British practice is to make these products with a bread flour, possibly a standard baker's grade flour or an even stronger flour. Some plant bakeries use a flour that is 75% Canadian Western Red Spring wheat to make croissants. A grist this strong has an enormous amount of tolerance.

Croissants are a high added value product so the cost of the strong grist can be recovered in the price. The very strong flour is a guarantee that the product will work despite any problems in the bakery.

The Process. A reasonable process to make this product would be to mix a straight dough with 10% sugar, 1% salt, and 4% margarine (or butter) with 9% yeast. All of these percentages are based on the weight of the flour. A sponge batter method could also be used.

The dough is mixed cold (18–20°C), which retards fermentation, requiring a high level of yeast to speed up the process. Next the dough is proved for up to 1 hour with a temperature not exceeding the melting point of the fat (around 37°C).

The dough is then sheeted out and the fat is spread on two-thirds of the dough. Usually there is one part of margarine or butter to four parts of dough by weight. The dough is then folded to give two layers of fat separating three layers of dough. The dough is then sheeted out and folded again. The dough is then rested at refrigerator temperature. Subsequently, the dough is sheeted out and cut into triangles with a height roughly twice the width of the base. The triangles are then rolled up into the crescent shape (Figure 16).

The product is then proved for 1–3 hours and is baked at around 230°C for 20 min. If the product is under proved the flakiness will be lost and a tough and chewy product results.

Figure 16 *A croissant*

7.4.7 Hot Cross Buns

These spiced buns with dried fruit in them are another product with a history. In the Middle Ages it was the custom to make buns with a cross on them. These buns assumed a mystical significance that caused the Cromwell's puritans to ban them, except for immediately before Easter. Small bakers generally keep to this rule while supermarket in store bakeries sell them for a much longer period of time.

Commercially these products are made from bread flour, either by a straight dough process or by a sponge batter process. The dough normally contains whole egg. The dried fruit is added as late as possible in the process, just before shaping and proving. If the product is made in a spiral mixer the low energy reverse setting is used to incorporate the fruit.

The cross can be made merely as a cut on the bun or, more likely, a paste of flour and water is piped onto the bun.

7.4.8 Buns

Although many of the products covered in this section are not made by the CBP process, buns can be if required. Possible recipes are shown in Table 4.

After initial development the dough is fermented for hour or an hour and a half at around 24–28°C. The dough is then knocked back three-quarters of the way through proving. Next the buns are moulded to size and proved in a cool prover with steam to prevent skinning. The products are then baked, possibly at 210–220°C.

If required, dried fruit would be folded in before moulding. Note the savings from the extra water and the saving from a softer flour achieved by the Chorleywood process.

Table 4 *Recipes for buns using either bulk fermentation or the Chorleywood process*

Bulk fermentation		*Chorleywood process*	
Parts by weight	*Ingredient*	*Parts by weight*	*Ingredient*
1000	Strong flour	1000	Less strong flour
70	Yeast	85	Yeast
30	Skimmed milk powder	30	Skimmed milk powder
125	Sugar	125	Sugar
125	Fat	125	Fat
75	Whole egg	75	Whole egg
10	Salt	10	Salt
500	Water	510	Water
0	Bread improver	10	Bread improver

7.4.9 Danish Pastries

This is another product whose origins are not what they seem. In Denmark they are known as Wienerbrot, indicating an origin in Vienna (Wien in German). They are made from a strong flour made into a rich, sweet, yeasted dough. Given that this is a long process a high Falling Number, *i.e.* low amylase activity is desirable.

The unusual feature of the process is that the dough is partially developed mechanically during the many rolling and sheeting processes. The rather soft dough is mixed cold around 16 to 18°C. The dough is then retarded for several hours, during which time some fermentation takes place. The cold dough is then sheeted out and the fat is rolled in as for croissants. The preferred fats are butter or a specialised margarine made for the purpose.

The dough sheet is returned to the retarder for 6–12 hours after being sheeted and folded twice. The dough is then sheeted and folded again followed by another 24–48 hours in the retarder.

The pastries are then made by sheeting out the dough and cutting shapes from it. Various sweet fillings are added such as nuts, cinnamon, fruit purée with or without creme patissier. After any further shaping the pieces are proved. The products may be glazed with diluted egg to make the finished piece shine or the same effect can be achieved by spraying the finished product with a light sugar syrup as it leaves the oven (Figure 17).

Danish pastries are a product that suit the small baker and the in-store bakery. They have become more popular in the UK in recent years. They have always been relatively more popular in the USA.

Figure 17 *Danish pastries*

7.4.10 Pretzels

Pretzels are a popular snack in the USA, Germany and some other European countries. There is a tradition of eating pretzels with beer in the USA and Germany. They are not particularly common in the UK but are becoming more so.

Pretzels are supposed to have originated in Italy in a monastery in the Alps around the year 600 AD. Allegedly, the monk running the bakery formed strips of bread dough into a shape that resembled arms folded across the breast as in prayer. These were called pretiola (which means little reward in Italian) and were given as a reward to those children who learned their catechism. The product travelled through the Alps to Germany, where the name became pretzel or bretzel.

These soft pretzels are bread and have a shelf life like bread. The only way of extending the shelf life is to reduce the water content to around 2–3%. They remain popular and are still made.

The hard pretzel is said to be a serendipitous discovery made when an apprentice baker left a tray of soft pretzels in the oven overnight. The resulting hard product was liked and started a whole new business.

The manufacture of pretzels is a large and very competitive business. All manufacturers claim that their product recipe and process is superior to their competitors. In view of this the following is a way in which these products can be made.

Pretzels are made from a very stiff dough that is made from a low protein soft wheat flour, *e.g.* one made from a soft white winter wheat. The water level would be only 38–42%, with 0.25% yeast and 1% shortening, 1% salt and 1% dry malt. The dough would be mixed in a "Z" blade mixer and then left to prove for up to 4 hours.

The dough is then divided into small pieces that are rolled in to a small strip that is then shaped into the pretzel shape. This shaping process is often carried out by extruding the dough through a special die to produce the pretzel shape. A straight die is often used to give the product in a straight stick form.

The shaped pretzels are then given a short intermediate proof followed by passage through an alkaline bath of baking soda. Next the pretzels are covered with salt and baked at 230°C for 4–5 min. The effect of the alkaline treatment is to produce a crisp shiny crust on the baked piece as a result of the reactions of the starch reducing sugars and protein.

The baked pretzels are then fed through a drying stage at 120°C for 0.5–1.5 hours, depending on the size of the piece. The reason for the length of the drying step is that the rate of drying is controlled by the rate of diffusion of water in the product. The finished product is then wrapped to keep the moisture level below 3.5% (which is the point when the product become unacceptable).

7.4.11 Not Baked

Some products that are similar to baked products are either fried or boiled instead. Strictly, these products are outside the scope of this work, but they are sufficiently similar to be included.

7.4.11.1 Doughnuts. Doughnuts are made by deep frying pieces of a lean sweet dough. The dough is mixed then fermented. Next it is sheeted out and cut into rings. These are then proved and for 0.5–0.75 hour before frying for 1 min on each side. The leavened doughnuts float on the surface of the fat. The doughnuts are next removed from the fryer and allowed to drain. Then they are glazed with a sugar water glaze or rolled in sugar.

Doughnuts are made in many shapes but the ring shape has acquired some folklore. It is said that a ship's captain had them made like this to allow them to be put on the spokes of the ship's wheel. There is a more prosaic explanation. If the product is fried the heat can only travel slowly through the product because it is a poor conductor of heat. The hole in the middle prevents a situation where the outside becomes over cooked before the middle is cooked.

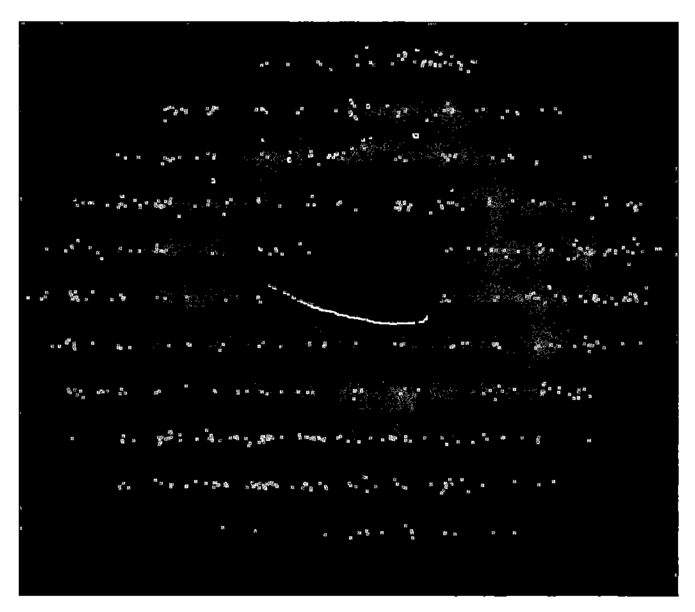

Figure 18 *A typical bagel*

7.4.11.2 Bagels. These are another ring shaped product (Figure 18). They also are said to originate from the time when Vienna was fighting off the Turks. It is said that the bagel was developed as a tribute to Jan Sobieski, king of Poland and a famous horse man. The ring shape is said to resemble a stirrup (buegel in German). The hard ring shape is obtained by immersing the ring of dough in boiling water.

Bagels are made traditionally by using a high protein hard wheat flour (around 13.5–14% protein). The dough is then treated with a combination of proving and retarding. The retarding can be done before proving or afterwards, with a time as short as 8 hours or as long as 24 hours. These steps are critical since an over-proofed bagel will collapse in the oven while an under retarded one will "ball up", *i.e.* expand and close the centre hole. A traditional bagel should have tiny blisters on the surface. These are caused by lactic acid formed during the retarding step. Alternatively 0.1–0.5% lactic acid can be added.

Immediately before baking the proved bagels are boiled by dropping them in boiling water. They float and are boiled for 1 min before being flipped over and boiled for 1 min on the other side. This process gelatinises the starch, giving a hard shiny crust when the bagel is baked. The product is then traditionally baked in a hearth oven although they can be baked in other ovens.

Poppy seeds and sesame seeds are sometimes added by applying them to the outside of the boiled bagel before baking. Other flavourings, *e.g.* onion, are mixed in to the dough.

Bagels are a product that is much more common in the USA than the UK. A few years ago they were unknown outside the Jewish community in the UK. They have become much more common in recent years. Newer varieties of bagel have appeared, made by adding eggs, making a rich dough product, adding fat or enzymes to make them softer or sweetening them by adding honey or raisins. In the UK, the use of a lower protein flour and pure dried vital wheat gluten might be attractive.

CHAPTER 8

Products Other than Bread

8.1 PUFF PASTRY

Puff pastry is one of the more difficult types of pastry to make. Good puff pastry should rise well and be crisp. Puff pastry is an example of a laminated product. The uncooked dough consists of layers of dough separated by fat. When the pastry is baked the water in the pastry turns to steam that pushes the layers of pastry apart (Figure 1). The heat gelatinises the starch, causing it to set on cooling.

8.1.1 Methods

The methods of making puff pastry can be divided into the Scotch method and the English or French method.

8.1.1.1 The Scotch Method. In this method the fat is cut into small cubes of about 20 mm sides and added to the mixing bowl with all the other ingredients.

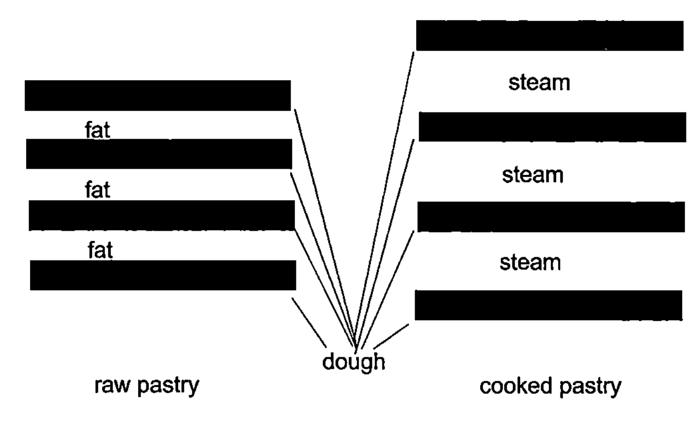

Figure 1 *Formation of puff pastry*

8.1.1.2 The English (also known as the French) Method. In contrast, in the English method the fat is rolled out between the layers of pastry. Not too surprisingly the Scotch method is less likely to give a discrete layer of fat between the layers.

8.1.1.3 A Comparison. The Scotch method does not give a continuous layer of fat between the dough layers; thus when baked the lift of the pastry will be less as the steam is more likely to escape. A low degree of lift and a flaky texture is a requirement in the pastry of a sausage roll so the Scotch method is often used here. An advantage of the Scotch method is that because there are not discrete layers of dough, but dough intermingled with fat, development of the gluten is less likely to take place; hence there is less need to rest the dough to relax it. The uncertain lift given by pastry made by the Scotch process means that pastry made is way is not suitable for products like vol-au-vent cases where dimensional stability is needed.

8.1.2 Type of Flour

Puff pastry is one of the few products that is made with both hard and soft wheat flour. Both types of flour have different advantages. Dough development of the sort that is desirable in bread is undesirable in puff pastry. As a dough develops the elasticity increases while the extensibility reduces. Extensibility is very desirable in a product that is made by making many thin layers of dough. The rolling and re-rolling of the dough causes mechanical development of the dough to occur (see Chapter 7 on bread). The traditional way of handling this is to rest the dough, *i.e.* leave it standing. A soft wheat flour will have less tendency to undergo dough development than a hard wheat flour and will need less resting time to be given to the dough. The disadvantage of soft wheat flour is that the dough may become too soft on standing if there is a delay in the plant. A hard wheat flour is more suited to a plant bakery. Probably the best choice for puff pastry would be an untreated bread flour, *i.e.* without the additives that encourage dough development. One advantage of the hard wheat flour is that the properties of this sort of flour are much more tightly controlled than soft wheat flour.

8.1.3 The Type of Fat

The fat present pays an important role in puff pastry since it has to separate the layers of pastry. To produce the most effective barrier a fat with the highest possible solid fat index is required, *i.e.* a very hard fat. Fat suppliers have developed special very hard margarines expressly for making puff pastry. The traditional fat for this purpose is butter, which is much softer. In certain European countries it is a strong marketing plus to label pastry as made with pure butter.

8.1.4 Additives

Various additives are used to act as dough relaxants. They are L-cysteine hydrochloride, sulfur dioxide, possibly in the form of sodium metabisulfite, and acids such as acetic acid in vinegar.

L-Cysteine hydrochloride presumably acts by breaking –S–S– links. Acetic acid in the form of white, *i.e.* distilled vinegar, is an old baker's method for relaxing pastry. Acetic acid if used would count as an additive while white vinegar is an ingredient. Presumably the acetic acid protonates the proteins and discourages their association. Sulfur dioxide acts as a reducing agent.

There are obviously savings to be made in using a dough relaxant rather than having dough sat waiting in the bakery to relax it.

8.1.5 Re-work

Any operation that involves cutting shapes out of pastry will inevitably accumulate off cuts. In domestic practice these off cuts are re-rolled and re-used. Re-rolling the off cuts in a bakery tends to be problematic. The act of re-rolling is likely to cause the dough to develop and there is the possibility of microbial contamination. The best solution is to use a pastry relaxant and to add the re-work back into the mixer so that it is distributed in the bulk of the pastry rather than allowing it to be re-rolled directly for re-shaping. Re-rolling and direct re-shaping will give a product that is 100% recycled and will likely behave differently from fresh product.

8.2 SHORT PASTRY

As this is a product where gluten development is not desirable, the flour usually used is a non-bread making flour such as an English soft wheat flour. The other typical ingredients are water, fat and possibly salt. Some pastry is made with self raising flour or by adding baking powder.

The traditional way of making these products is to mix the fat and the flour, a process known as rubbing in the fat. This name is understandable, deriving from the time when the fat was literally rubbed into the flour with the finger tips – this has the effect of water proofing the flour, which reduces gluten formation as the proteins need to absorb water for

Table 1 *A recipe for short pastry*

Parts by weight	*Ingredient*
1000	Soft flour
10	Baking powder
250	Shortening
250	Margarine
125	Caster sugar
125	Water

the dough to develop. An old baker's saying is that cold hands were needed to make pastry. This was because the fat needed to be in the solid phase.

Margarine suppliers have promoted the idea of using a soft margarine and mixing the pastry in one step. This has proved popular with home bakers.

Some gluten development is desirable as otherwise the pastry will not be cohesive during sheeting and can crack. If there is excessive dough development the pastry will be too cohesive and will become tough and unpalatable. As often happens in bakery products a certain amount of development is good but too much is bad.

Table 1 gives a possible recipe for this sort of pastry.

8.3 HOT WATER PASTRY

This is the type of pastry traditionally used to make products like pork pies. As these are savoury products there is no sugar in the pastry.

These products are also known as raised pies, which indicates the problem in their manufacture. The pastry case has to be shaped and stand up while the filling is added. Presumably, a long time ago, someone considered the problem of making a raised pastry case. If the case was made of ordinary pastry it would collapse on standing. One way of preventing the case collapsing would be to bake it first filled with some inert material. This scheme would work but when the case had been filled with meat the pie would need to be cooked and the case, after being cooked twice, would be inedible.

The solution is to use hot or boiling water in mixing the pastry. This partially gelatinises the starch in the flour as well as melting the fat. A hard fat is used in this sort of product, either lard, the traditional material, or a hard vegetable fat. The hot water could also reduce the activity of some of the enzymes present such as amylase.

The melted fat is much more easily mixed into the pastry than it would be if the pastry was mixed cold. This intimate mixture gives the pastry a coating of fat, making it less likely to soften by picking up water from the filling if the filling has a lower water activity than the pastry. The shelf life of the product is then limited by the filling rather than the pastry.

The Melton Mowbray pork pie has recently been awarded defined origin status by the EU.

8.4 SCIENCE OF BISCUITS

The name biscuit is said to be derived from the Latin *panis biscoctus* meaning twice cooked bread.

The product that this referred to was ship's biscuits. These were made by cooking a dough and then drying out the product.

The drying out was the only available way of lowering the moisture content sufficiently to preserve the product. As the product consisted of just flour and water, cooked and dried, the end result must have been unpleasant.

The term biscuit now covers a whole range of products, even including pet foods. The crisp savoury products known as crackers are considered as biscuits while wafers are not biscuits.

This is an area where English and American usage have parted company. American usage has developed the word cookie, presumably from the Dutch Koekje for a small cake.

The manufacture of biscuits is an area where the plant bakeries produce most of the output. A few biscuits are made domestically, some are made by small bakers and supermarket in-store bakeries but these are only producing a fraction of the plant bakeries' output.

As a long life food biscuits were an early item of international trade. It is recorded that English biscuits were being exported to the USA in the nineteenth century.

8.4.1 Flour for Biscuits

It remains a useful rule of thumb that the properties of a good biscuit flour and a good bread flour are opposite. Bread flours are made from hard wheat with a high protein content, high starch damage and high Hagberg Falling Number. Biscuit flours are made from low protein content, low starch damage flours where a high Hagberg Falling Number is not very important but a moderately low Falling Number can be helpful.

The most important dough property in biscuit doughs is extensibility. Resistance is undesirable. The only type of flour that can not be used to make sweet biscuits is the sort of high protein hard wheat bread flour that is preferred for bread. Biscuits made from this sort of flour have a bread-like texture.

There are some soft wheats, both French and North American, that have bread making properties. Flour made from these wheats can be used for biscuits.

The other difference between bread flour and biscuit flour is in flour treatments. Bread flour treatments are oxidising agents while biscuit flours are treated with reducing agents. Although ascorbic acid, which is chemically a reducing agent, is used as a bread flour treatment it is a reaction product from the ascorbic acid that produces the oxidising effect.

The normal flour treatment for biscuit flours is sulfur dioxide. This can either be applied in the gaseous form in the flour mill or sodium metabisulfite can be added to the dough. The sodium metabisulfite releases sulfur dioxide on wetting and heating.

The preference for low protein, low starch damage flour in biscuits is obvious when the role of protein and damaged starch as water binders is considered. The aim in making biscuits is to produce a low moisture food. Incorporating components that bind water makes that aim more difficult. Table 2 considers the relative properties of biscuit and bread flours.

8.4.2 Fats

Most types of biscuits contain a considerable amount of fat. The traditional fat for biscuits was butter and some all-butter biscuits are made. Butter in this case brings flavour and a marketing plus to the product.

As biscuits are a long-life product any fat used in them has to be stable under the conditions of storage. Fats can deteriorate in two ways. Firstly, any fat can suffer lipolysis if the fatty acids are split from the glycerol. This is what happens in soap making, where the fatty acids are

Table 2 *Comparison of flours used bread and biscuits*

Property	*Bread flour*	*Biscuit flour*
Protein	High	Low
Starch damage	High	Low
Dough extensibility	Low	High
Dough resistance	High	Low
Wheat	Hard	Soft
Flour treatment	Oxidising agents	Reducing agents

chemically split from the glycerol. This is normally achieved by boiling the fat with caustic soda. Obviously, such a process would completely destroy a biscuit and indicates that chemical lipolysis is not going to happen during storage under any normal conditions.

Unfortunately, there is another way in which lipolysis can occur and that is under the influence of fat-splitting enzymes known as lipases. These enzymes have several sources: they are present in mammals, bacteria, moulds and cereals. Enzymic lipolysis can be prevented, in principle, by removing all sources of lipases, inactivating lipases by heat or reducing the water activity to a point where the lipase can not work. Biscuits rely on the effect of heat in the oven and the water activity of the product. The effect of lipolysis depends on the fatty acid present. In the case of butter which contains butyric acid a small increase in the amount of free butyric acid actually increases the buttery flavour. With lauric acid, which is present in hardened palm kernel oil, even low levels of the free acid give a taste of soap. Hardened palm kernel oil has been use in biscuits in applications such as the "cream filling" between two biscuits without problems, indicating that no lipase activity is taking place.

The second way in which fats deteriorate is oxidative lipolysis. This is an entirely different process in which oxygen free radicals add across double bonds. Oxidative rancidity can be prevented or reduced by several different routes. One way is to ensure that no double bonds are present. Another is to use anti-oxidants that act as free radical traps. Exposure to oxygen and ultraviolet light should be avoided. Reducing the temperature has no effect since free radical processes have a zero activation energy.

Oxidative rancidity is not necessarily a problem unless a polyunsaturated fat such as sunflower oil has been used. Where such a fat is used oxidative rancidity can occur and, as autoxidation occurs where the reaction becomes self-catalysing as oxygen free radicals react, some very unpleasant tastes can appear very quickly. In general, the effect of oxidation on biscuits at the end of their shelf life is that a cardboard taste starts to appear.

In the past various marine oils were used in biscuits but this practice had now ceased. Most biscuits contain vegetable fats. It was common to use hydrogenated fat in biscuits because vegetable oils were too soft physically and too unsaturated to be stable against oxidation. If an unsaturated fat is hydrogenated until it is completely saturated then a saturated fat results; however, if it is only partially hydrogenated a partially hydrogenated fat with several double bonds is produced.

The process of hydrogenation produces trans double bonds. Now trans double bonds are rare in nature and have been identified as unhealthy. This has led some countries to limit trans fatty acids in foods and has caused biscuit manufacturers to move away from partially hydrogenated vegetable fats. The fat suppliers have had to move from hydrogenation to fractionation to obtain the desired properties in their products.

Biscuits have been identified as a major source of dietary fat in the UK, being some 4% of that fat.[1] Notably, biscuits are not the source of the remaining 96%. Fats are not added to biscuits for no reason, they are there because they impart a certain texture. When the dough is mixed, if the flour has been coated with fat the fat prevents a gluten network being formed. This leads to a less hard product with a shorter more melt in the mouth texture. A high fat dough breaks apart easily if stretched, *i.e.* it is short and hence fats are referred to as shortening.

If appropriate emulsifiers are used it is sometimes possible to reduce the fat content of a recipe by about 20%. This change could be induced either by nutritional considerations or by a desire to reduce the use of the most expensive major ingredient.

Apart from butter or butter oil most fats that are used in biscuits are defined in terms of their physical and chemical properties. Fat suppliers are skilled at producing products with controlled physical and chemical properties from a range of raw materials. The baker can either buy fat on a physical and chemical specification, *e.g.* solid fat index, slip melting point, and not to contain lauric fat, or on an origin basis, *e.g.* to be coconut oil. The advantage of the botanical specification is that the item is a commodity and can be obtained from numerous sources. The disadvantage of this approach is that the product is tailored for a particular use.

When butter or margarine are used they are normally handled at temperatures near 18°C, which leaves the ingredient manageable but without breaking the emulsion. Specially blended dough fats can be handled at this temperature. Alternatively, the dough fat is handled at temperatures near 27°C, where it is pumpable. It is possible to use these fats at up to 40°C.

8.4.3 Sugars

Apart from the obvious function of adding sweetness, sugars affect the structure of biscuits. Biscuits made from short doughs contain the most sugar while semi-sweet biscuits contain less and crackers least. The water content of short doughs is so low that the sucrose present can not dissolve. Thus the crystal size is important as the sugar will be present in the solid form.

If there is a high level of sucrose in a short dough a hard glassy biscuit will be obtained, presumably because the molten sucrose forms a glass on cooling. The presence of glucose syrup will soften such a biscuit.

All reducing sugars present can undergo the Maillard reaction and produce attractive colour and flavours. Of the possible sugars present dextrose, fructose, maltose and lactose are all reducing sugars. The only non-reducing sugar likely to be present is sucrose.

The usual sugar ingredients in biscuits are sucrose, usually as a solid, and glucose syrup. Lactose can be present directly as an ingredient or in skimmed milk or whey powder. Another sugar-containing ingredient is malt extract, which contains maltose and adds both flavour and colour. Malt extract is a way of getting sugars into the product without them appearing as sugars on the label. The sugars would be listed of course in a nutritional statement.

Malt extracts for biscuits are chosen according to the level of colour required. One standard item of the specification would be that the extract should be non-diastatic, *i.e.* that the starch splitting enzymes of the original malt have been inactivated.

Maltodextrins, which are effectively a low DE glucose spray-dried syrup, are sometimes added to biscuits to enhance crispness. The low sweetness of the maltodextrins is a plus since they can be used in savoury biscuits where the sweetness of sugar would be inappropriate.

It is possible to make biscuits where the sugars have been replaced by sugar-free bulk sweeteners. Isomalt is particularly successful as a sucrose replacer in this application.

8.4.4 Milk and Other Dairy Ingredients

While liquid milk is little used in biscuit manufacture for practical reasons to do with lack of stability, skimmed milk solids are used. The preferred ingredient is skimmed milk powder. This is normally dispersed in twice its own weight of water to ensure that it is evenly dispersed in the finished product. The reconstituted milk powder has the same keeping properties as liquid milk so it must be refrigerated. Merely dry blending the milk powder is likely to produce a finished product with small brown specks of caramelised milk powder in it.

Milk powder contains several useful components, namely protein and lactose. Lactose is a reducing sugar that undergoes the Maillard reaction to produce flavour and colour. The proteins as well as participating in the Maillard reaction have useful emulsifying abilities. These benefits are only obtained if the lactose is dissolved and the proteins dissolved or dispersed.

Despite the above the use of milk powder in biscuits is reducing in the EU. The reason is the price of milk powder, which is dictated by the common agricultural policy. Some old recipes were formulated when milk powder was relatively inexpensive and, hence, had more milk powder than was needed. These recipes were the first to have their milk powder content reduced.

Another reason for the reduced use of milk powder is the availability of substitute ingredients such as lactose, whey powder or syrups and speciality milk powder replacers. The speciality milk powder replacers are produced by the dairy industry from milk or whey components, combining them to produce a product that will act as a substitute for milk powder but is less expensive.

Whey powder can be used as a substitute for milk powder but the flavour effects of some whey powders do not suit all biscuits. In the case of cheese flavoured savoury biscuits whey powder can add useful flavours.

The major component of whey is lactose, and while there various uses for the protein in making protein concentrates these leave a surplus of lactose. Impure grades of lactose have been available to the food industry for some time. They are relatively successful in biscuits as a raw material for the Maillard reaction to produce pleasant colours and flavours.

One of the bars to the use of lactose as a food ingredient is its limited solubility. This can be overcome by enzymatically splitting the lactose to its component monosaccharides dextrose and galactose. These monosaccharides are much more soluble than lactose but can still undergo the Maillard reaction. If the lactose is split into its constituent monosaccharides, whey can be condensed to a relatively stable high solids syrup. These syrups tend to carry some cheese flavour notes but, where used appropriately, can contribute beneficial colours and flavours in biscuits.

A dairy ingredient that is often used in biscuits is butter (see Section 8.4.2 on fats).

8.4.5 Other Cereal Ingredients

A few other cereal-based ingredients go into some biscuits. The most important is oats in the form of oatmeal or oat bran. Health claims are sometimes made regarding oat bran products. Some savoury biscuits have whole or kibbled grains of either wheat barley or oats. Barley does go into biscuits in the form of malt extract. Brewers or distillers spent grains, *i.e.* the insoluble fibres left after the soluble carbohydrate has been removed, have been suggested as a high fibre ingredient in biscuits.

8.4.6 Mixing Biscuits

Biscuits can be mixed in most sorts of mixer. Continuous and batch mixers are both used. Some of the batch mixers are vertical others are horizontal. It might be thought that, as biscuits are often shaped and baked continuously, a continuous mixer would be essential. In practice, dough can always be mixed faster than it can be used, so a continuous mixer is not essential.

However, continuous mixers do exist and are used. The problems are that they are difficult to start up, continuous metering of ingredients is expensive and they have to be shut down in the event of a breakdown further down the line.

While all biscuit doughs need to be mixed the other requirements of the mixer depend on the type of dough involved. Biscuit doughs are normally classified as hard developed doughs, semi-sweet doughs, short doughs and batters. The needs of each type are considered separately below.

8.4.7 Types of Dough

8.4.7.1 Hard Developed Doughs. These doughs are used to make crackers. The mixing action has to develop the dough as in bread. Indeed some crackers are fermented with yeast like bread. Crackers are made from a dough that is low in fat and sugar but relatively high in water.

Cracker doughs are mixed in an all in one process that involves kneading the dough to develop the gluten. The dough would be mixed to a final temperature of 26–30°C, which is obtained by controlling the energy input and the temperature of the ingredients. The dough is then left to ferment. After fermentation some cracker doughs are remixed with more flour and water.

8.4.7.2 Semi-sweet Doughs. These contain more sugar and fat than crackers. Mixing should be to 41°C if sodium metabisulfite is used and to 45°C if it is not used. Mixing time is not critical. Semi-sweet doughs are normally mixed on an all-in-one basis.

8.4.7.3 Short Doughs. Gluten development is not desirable in these doughs. The high level of fat and sugar preclude hydration of the gluten. The problem is to disperse the fat and water throughout the dough without developing the gluten. The level of sugar is so high in these products that it can not all dissolve in the water.

These doughs are mixed in a two-stage process by forming an emulsion of the fat in the water and then adding the flour. Energy input can be high in the first stage as this helps the dispersion and there is no gluten present to develop. The second stage mixing, where the flour is added, is very short to avoid developing the gluten. In some cases some of the sugar is added with the flour.

The problems then are making a satisfactory emulsion in the first stage but avoiding gluten development in the second.

8.4.7.4 Batters. A few biscuits are made from a dough so soft that it is really a batter. These products sometimes contain eggs. As these products are nearly cakes they are made in a cake-type mixer with a high sheer rate to incorporate air.

8.4.8 Shaping Biscuits

If biscuits are made by hand the shaping process would be to roll out the dough and use a cutter to cut the biscuits to shape. The scrap dough is then re-rolled and more pieces are cut with the excess being re-rolled and the process repeated until there is insufficient dough to make any more biscuits.

It is also possible to shape biscuits by a mechanised system that does the same process. This is called sheeting gauging and cutting. Some biscuits are shaped by extrusion and depositing, while others are wire cut. Alternatively, the biscuit dough can be fed to a rotary moulder, which shapes the biscuits in one step (Figure 2).

8.4.8.1 Sheeting Gauging and Cutting. Sheeting is the equivalent of the domestic rolling pin. The dough is fed from a hopper between rollers

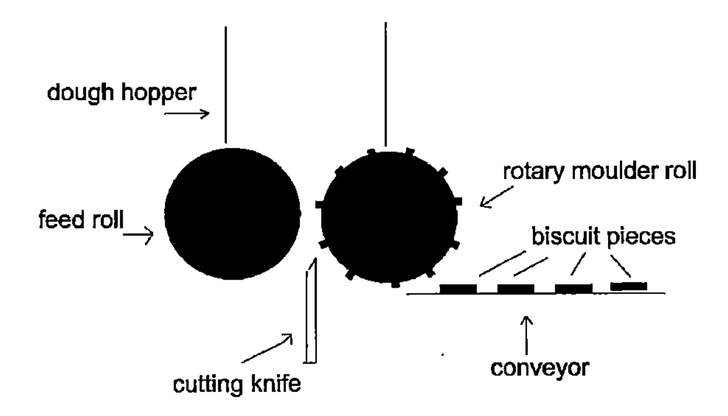

Figure 2 *Sketch of a biscuit rotary moulder*

to produce a dough of controlled thickness. The problem with this process is that the rolling works the dough, which causes the gluten to develop. This problem can be minimised by using only one set of rollers. The rollers that determine the thickness of the dough are known as gauge rollers. If the dough is fed to the cutter under tension the dough pieces will shrink during cooking, tending to emerge thicker at the front and back.

The answer to this problem is to allow the dough to relax before cutting. In some older cutting machines a very long conveyor was provided that led to the cutter. The dough was over fed to this conveyor from the final gauge roll, resulting in ripples. In a properly set up machine the ripples are absorbed by the dough shrinking, presenting a smooth sheet of dough to the cutter.

To avoid the very long conveyor needed in the such systems, more modern plants use an intermediate conveyor. The dough is overfed to this web, forming ripples. If the speed of this belt is adjusted correctly the dough can be presented to the cutting web without ripples.

In older machines the dough then passes to the cutting web where the cutters come down and cut the biscuits to shape. The cut pieces are then fed to the oven. With a travelling oven this is just a transfer from the cutting band to the oven band. In this sort of machine the heavy block cutters drop on to the cutting web, move with the dough then rise, swing back and drop again. This type of cutter is a mechanised version of a hand cutter.

Problems with the limited maximum speed and the high maintenance requirement led to the development of rotary cutting machines. In these systems the dough is fed between either one or two pairs of rolls (Figures 3 and 4). In either system the lower of the pair of rolls acts as an anvil while the upper roller squeezes the dough, dockers it, pins it to the web and cuts the outline. Provided this works it is the cheaper of the two solutions. Problems occur when the dough lifts from the cutting belt as the two operations of pinning and cutting can not be adjusted independently.

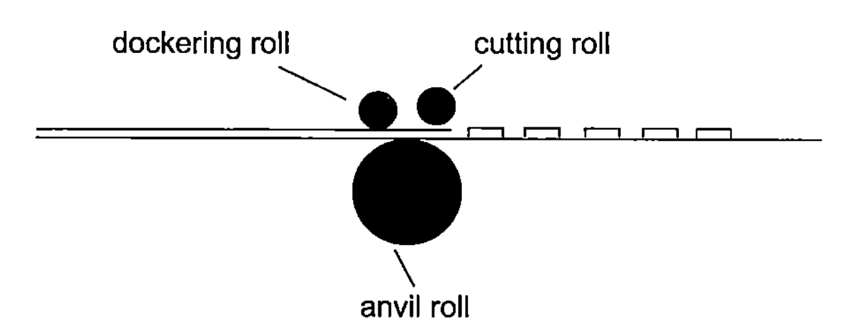

Figure 3 *A single anvil biscuit cutting machine*

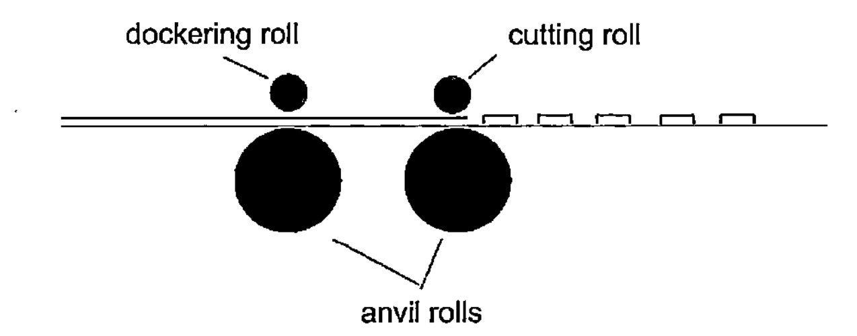

Figure 4 *A double anvil biscuit cutting machine*

In the system where two pairs of rollers are used the first pair dockers the dough, pins the dough to the cutting web and prints any surface pattern or writing. The second pair of rollers cuts out the biscuits leaving the usual pattern of scrap dough. Thus there is a choice between a cheaper solution that sometimes works and a more expensive solution that will work under a wider range of conditions.

The problems at the cutting stage are that the dough must stick to the cutting web and not the cutter. The scrap dough has to be recovered and remixed into the system in such a way that the performance of the system is not impaired.

8.4.8.2 Extruding Wire Cutting and Depositing. The simplest way of shaping biscuits is to extrude them through a die. If the dough is very soft, effectively a batter, it can be deposited. If the dough is firm it can be wire cut. The dough is forced through a die and a taut wire held in a frame moves across the die, cutting the pieces. The wire is moved away from the die for the reverse stroke as extrusion is continuous. The wire can be vibrated as it cuts or it can be replaced by a straight or serrated knife blade. Any of these cutting tools can be coated with polytetrafluoroethylene (PTFE) to reduce sticking. Wire cutting can handle sticky doughs and doughs with inclusions such as nuts or oatflakes that could not be rotary moulded.

Very soft doughs that are not suitable for rotary moulding are extruded onto a belt to produce strips of dough. These dough strips can then be guillotined before or after baking.

8.4.8.3 Rotary Moulding. Rotary moulding is one of the two main ways of shaping biscuits. It is used for producing biscuits from short doughs. The dough is forced into an impression in a roller, the excess dough is scraped off by a fixed knife blade, and the shaped biscuits are carried away on a fabric belt.

Rotary moulding has several advantages over wire cutting. It is easy to produce complex shapes simply by making the impression the negative image of the desired shape. Thus, any writing has to be mirror writing.

The major advantage is that there is no re-worked dough being recycled from a cutting unit. The problem with re-worked dough is that it becomes tougher than fresh dough.

8.4.9 Baking Biscuits

While biscuits can be and are baked in almost any type of oven, including deck ovens, rack ovens and travelling ovens, most biscuits are baked in travelling ovens. These ovens suit large plant bakeries. The throughput of these systems is measured in terms of kilos per hour. One of their advantages is that it is possible to arrange the oven in a series of zones so that the product passes first into the hottest part of the oven and is moved to cooler regions as cooking proceeds. This is what would happen with a clay oven fired by filling it with wood. In a tunnel oven it is sometimes best to arrange the zones so that the first zone is less hot than the second zone. This prevents the surface of the biscuit becoming too hard too quickly, which could produce a case hardened layer that resists the removal of moisture.

8.4.9.1 Chemical and Physical Changes in Baking a Biscuit. The first thing that will happen is that any gases, including air and carbon dioxide from leavening agents, will expand, causing the biscuit to expand. Water will be converted into steam, also causing the biscuit to expand. This expansion is the oven spring.

The Maillard reaction will take place between the proteins and reducing sugars on the surface. As the interior of the biscuit heats up by conduction this reaction will spread to the interior. If the biscuit is excessively alkaline from too much sodium bicarbonate a yellow colour will be produced.

The proteins present will start to denature and the starch will start to gelatinise. These processes cause the structure to set.

Throughout the cooking water is lost from the biscuits. At the end of the cooking process this is the only change that happens. The loss of water will be controlled by the rate of diffusion of water from the middle to the surface. It is at this point that the dockering, *i.e.* the holes in the biscuit, assist the water removal.

Some biscuit makers have started to use a radiofrequency hood to remove a little extra moisture. This has the effect of making the product crisper and increasing the shelf life since the biscuit can absorb more moisture without becoming soft.

A further refinement is to dispense with the radiofrequency hood and use the sort of stoving system traditionally used to process confectionery gum products. The biscuits are treated by blowing heated air that has been dehumidified by use of a heat pump. This system uses markedly less energy than the radiofrequency hood.

8.4.10 Packaging

Biscuits need to be protected against breakage and the uptake of water. The very crispness of biscuits makes them susceptible to breakage in transit. The traditional packing for biscuits other than those sold loose is a tin. At one time the tins were returned, cleaned, and refilled. This practice seems to have been discontinued.

Tins of biscuits, particularly with a decorative tin, are a popular gift and souvenir item. Some years ago HM Customs and Excise attempted to claim that while the biscuits were a food the decorated tin was not a mere packaging but a separate supply and should pay VAT at the standard rate. This was a serious issue for the retailer concerned since Customs and Excise were demanding the VAT that the retailer had not collected! Eventually, the retailer won and the tins were ruled to be packaging.

Except in the case of biscuits in a decorative tin the manufacture is in a dilemma between the customer who dislikes broken or soggy biscuits, environmentalists who dislike excess packaging, and the finance director who dislikes excess spending on packaging.

One positive factor in biscuit packaging is that there are now available a whole range of polymers that act as both a moisture and a vapour barrier. These films offer much better protection than the waxed paper that was once used. The other improvement is the use of high speed highly mechanised packing machinery.

8.5 SCIENCE OF WAFERS

Wafers are an unusual product. They are often incorrectly included with biscuits, possibly because they are both made from soft wheat flour. Wafers, unlike biscuits, are a low fat, low sugar product. They normally consist almost entirely of flour. There is a very wide difference between the various sorts of wafers. Some wafers are made to serve with ice cream others are made to covered in chocolate and sold as confectionery.

All wafers have in common that they are baked to a low moisture content, as are biscuits. Wafers readily pick up moisture at UK ambient conditions and become soft unless they are protected from moisture. Wafers will normally soften more rapidly than biscuits.

8.5.1 Raising Agents

Wafers can be either yeast raised or chemically raised, *e.g.* by baking powder. Yeast raised wafers do not contain much sugar after fermentation.

8.5.2 Flour for Wafers

Wafers are normally made from a low to medium protein soft wheat flour. Too high a protein flour produces too hard a wafer. Conversely, too low a protein content will give very fragile wafers.

Depending on the colour and use of the finished wafer some wafers are made from brown flours. While the increased fibre content has nutritional advantages the real reason for this is that the higher extraction rate makes the flour cheaper to produce.

8.5.3 Production Process

Obviously the process is slightly different depending on whether the wafer is yeast or chemically raised. In essence the ingredients are mixed into a batter then baked. There is no need to develop the gluten, indeed gluten development is unhelpful. One particular problem in wafer batters is the separation of strings of gluten. This can be avoided either by mixing the batter cold or by careful selection of the flour used.

If the product is handmade on a small scale the flour used is less critical than if the product is made on a large scale on an automated plant. Mixing cold and baking the product before the gluten separates may be a useful small scale solution but on a large scale where the batter has to be pumped to the ovens to be baked it will not work.

The large-scale plant needs flour that does not undergo gluten separation. Conventional flour specifications give no information about any tendency to gluten separation. Careful Extensograph testing of successful and unsuccessful flours might give useful information.

What is actually needed is a test that rapidly differentiates between suitable and unsuitable flours. Making a test batter and mixing it on maximum speed in a food mixer is a fairly effective way of detecting those flours that will undergo gluten separation from those that will not.

8.5.3.1 Baking Wafers. On a small scale, wafers are baked by pouring the batter on a heated metal plate and bringing a second plate that is hinged down on the first, trapping the batter between the two plates. The plates are likely to have been treated with a releasing agent and may have a pattern inscribed on them. The plates will be equipped with a system for venting the steam produced in cooking. The heat is then applied and the wafer cooks very quickly.

Large-scale wafer ovens are essentially the small-scale system scaled up. One problem with the mass production of some wafers is that they are too delicate to use mechanical handling and must be moved manually.

8.5.4 Maturing Wafers

Wafers change after they are baked, both dimensionally and in terms of texture. It might be thought that this is either loss or gain of water to or from the environment but the same changes go on even if the wafer is hermetically sealed from the environment. The concept of free and bound water was used at one time to explain these changes. The water appeared to exist in two forms, one that could be removed and another that could not. This useful concept has since been discouraged since it has emerged, as a result of isotope tracer studies, that the water is interchangeable between the two states. NMR studies have revealed that there is a fraction of the water in these products that does not change its NMR signal on going from the liquid to the frozen state when the product is cooled to freezing temperature. The term non-freezable water has now replaced bound water.

Maturing is important practically because the wafers change dimensionally. If wafers are just cooled and covered with chocolate they will subsequently crack the chocolate. This can be avoided by first maturing the wafers.

8.6 CAKES

8.6.1 Introduction

Cakes more than any other category of product in this work vary in shelf life. They vary from sponge cakes to the sort of rich fruit cake that is used as a Christmas cake. Indeed it is difficult to define a cake except in terms of negatives. A cake is not yeast raised it is not a biscuit or pastry. Incidentally, Marie Antoinette probably said "why do they not eat Brioche". Brioche is a sweet yeast raised product and so it is not a cake!

8.6.2 Shelf Life

Cakes are in principle subject to all the threats to a long shelf life that any other bakery product is subject to. The product can dry out, the starch can retrograde or mould can grow. These are in addition to the threats of oxidation, loss of flavour and lipolysis by any enzymes present.

Enzymic action is not normally a problem since only the most heat resistant enzymes survive the cooking. Oxidation can occur, but unless some polyunsaturated material, *e.g.* sunflower oil, is present is unlikely to be the problem that limits shelf life. Products that have suffered oxidation tend in general to have a cardboard taste rather than any other effect.

Drying out can be prevented in one of two ways, either by packaging or by lowering the water activity of the product. Starch retrogradation can be inhibited by using a starch complexing emulsifier with or without the addition of fat. Mould growth is inhibited by a low water activity.

Notably, low water activity only inhibits mould growth. Most bakery products leave the oven in a sterile condition, any mould contamination is subsequent to baking. The effect of a low water activity is that the osmotic pressure is such that no mould spores can reproduce. Thus the product is safe unless the water activity rises locally.

8.6.3 Rich Fruit Cakes

These solve the shelf life problem by having a very low water activity aided by the quantity of dried fruit that they contain. The fat content and the low water activity inhibit starch retrogradation.

Mould can form on the surface of this sort of cake if it is placed in a container or wrapped before the cake has cooled completely. Under these conditions condensation can form on the container or wrapping. The concentration then drips onto the cake, causing a small spot of high water activity where mould can grow.

This type of cake is often stored for a while after baking because it is to be covered with royal icing. The cake not only needs to cool but the surface needs to equilibrate with the bulk. There is a paradox that if the cake is just wrapped in greaseproof paper and kept in a tin that is not air tight there will be no problem. However, if the cake is wrapped in aluminium foil and kept in an air tight tin there is a possibility of mould formation because with the better moisture barrier any condensation or sweating can not evaporate.

There is a practice of making holes in this sort of cake and pouring in spirits such as whisky, brandy or rum. While this may be done to enhance the flavour it will almost certainly improve the keeping properties. The alcoholic mixture will not only reduce the water activity as ethanol has considerable mould inhibitory and antibacterial properties.

8.6.4 Long-life Sponge Cakes

While an ordinary sponge cake has a life of a few days, various products consisting of a small individually wrapped sponge cake have appeared on the market. The fact that these products were covered in chocolate aroused some interest in the British confectionery industry. It remains a quirk of the British taxation system that while most food is zero rated for VAT, confectionery, ice cream and potato crisps are standard rated. While chocolate covered biscuits are classified as confectionery and are liable for VAT, chocolate covered cakes are classified as food and are zero rated. After very lengthy legal proceedings it was decided that "Jaffa cakes" are cakes and not biscuits. The benefit of VAT reduction is reduced by the fact that sponge cakes have to be made from relatively expensive ingredients such as eggs.

If an ordinary sponge cake was wrapped up and stored, while the wrapping would prevent it drying out the cake would quickly become mouldy. Wrapping these products is one of the easiest parts of the problem as a flow pack with a good moisture and vapour barrier seems to work well. If the shelf life is to be further extended then the water activity needs to be reduced. Since this is a colligative property this can be achieved by increasing the number of soluble particles. Ingredients that are used to reduce the water activity include dextrose monohydrate, high DE glucose syrup, invert sugar, golden syrup, sorbitol and glycerol.

The problem of starch retrogradation is tackled by using a starch complexing emulsifier. When all these problems have been attended to the shelf life can be measured in months rather than days.

The confectionery industry's interest in this sort of product as a count line seems to have diminished. The products are more likely to be found in multi-packs on the grocery shelves of the supermarket, aimed at the lunch box market.

8.6.5 Making Sponge Cakes

A sponge cake is made by first making a stable foam and then cooking it. There are several variations possible in terms of recipe and the method. Some recipes have fat in, others do not. The main push is to

Table 3 *Two sponge cake recipes*[a]

Ingredient	*Parts by weight for the original recipe*	*Parts by weight for the economy version*
Soft flour	1000	1000
Sugar	1000	1000
Whole egg	1000	400
Milk	0	700
Baking powder	0	50

[a] In either case 5% of glycerine can be added to improve shelf life.

reduce costs by minimising the use of eggs, which are the most expensive ingredient. However, eggs are a very useful ingredient as they stabilise the foam and set on cooking.

8.6.5.1 Traditional Sponge Cakes. An old fashioned recipe would be referred to as pint, pound, pound, *i.e.* a pint (571 mL) of egg, a pound (454 g) of sugar and a pound (454 g) of flour. The sugar would be caster sugar and the flour soft flour.

The warmed sugar (32°C) would be whisked into a stiff batter with the egg, then the flour is blended in carefully. The batter can then be piped into tins and baked at 204°C. As this is a fat-less sponge care needs to be taken to keep all the equipment fat free lest the fat cause the foam to break.

Some economies can be made by using less egg and some baking powder. Table 3 compares the two recipes.

Alternatively an "all in" method can be used where all the ingredients are mixed together in a power whisk.

8.6.6 A Comparison of Cake Making Methods

When two methods of doing something continue to be in use it usually works out that each has it particular merits or problem but neither is superior to the other overall. This is the case with the sugar batter and the flour batter methods of making cakes.

8.6.6.1 The Sugar Batter Method. This method works by making an emulsion of oil in water with air bubbles in the oil phase. All the other raw materials are dispersed or dissolved in the aqueous phase.

Initially the fats and sugar are mixed together to produce a light textured mixture. This process is known as creaming. The time this takes is very dependent on the mixer used and the form of the fat, *e.g.* solid fat or bakery margarine, but 10 min usually suffices. The batter can be controlled by mixing to a specified specific gravity.

Next the liquid egg is added in small portions. The egg needs to be added cautiously lest the batter should curdle. The egg might be added over 5–10 min in five or six portions. It is extremely important that the egg is at ambient temperature and not at refrigerator temperature. Dumping in a cold mass of egg is likely to cause curdling, *i.e.* the oil in water emulsion breaks down. All the ingredients should be at ambient temperature but only the egg and possibly the fat are likely to be kept in a refrigerator.

A well-made batter should have a velvety appearance and a smooth texture. Any further flour or other ingredients such as flavours are then mixed in gently. If dried fruit is being used this should be the last thing added.

8.6.6.2 The Flour Batter Method of Cake Making. This method is more similar to the method used for sponge cakes. The flour is split into two portions, one portion is creamed with the fat while the other one is held back to be mixed into the batter later. The eggs and sugar are whisked together to produce a foam as in a sponge cake. This can be done at the same time on a separate mixer if one is available or by using another bowl and beater on the same mixer.

The ratio of flour to fat used in the batter is normally 400 of flour to 450 of fat, *i.e.* slightly less flour than fat, while the whole egg is whisked with its own weight of sugar. Care needs to be taken not to beat too much air into the sugar and egg lest the resulting cake should be too light. This is not a problem in the sugar batter method as the presence of fat limits the amount of air that can be incorporated. Any minor ingredients such as flavours can be added to the flour and fat batter with colours if used.

Next the egg and sugar foam is carefully added to the batter in small portions or in a continuous stream. In either case the mixer should be running slowly. When the two components have been mixed any remaining flour can be added, followed the dried fruit if used.

If milk is added it should be added with the second portion of flour. Any remaining sugar above the weight of the egg is incorporated at this point, dissolved in the milk with colours and any salt that is added. If milk powder is used to replace liquid milk it is added along with the second flour addition in conjunction with sufficient water to compensate for that absent from the powder.

The flour batter method reduces the possibility of curdling and also gives a more even texture to the finished product. The flour batter method is also faster with a reduced risk of developing the gluten by excessive mixing. Gluten development is undesirable since it would render the batter and the finished product tough. The disadvantage of the flour batter method is that it is possible to produce too light a texture.

8.7 MISCELLANEOUS CHEMICALLY LEAVENED PRODUCTS

8.7.1 Doughnuts

Some doughnuts are made with a chemical leaven rather than yeast. Such doughnuts are sometimes known as cake doughnuts. The batter can be regarded as a leaner version of a cake batter but with less sugar. These products are often made from a pre-prepared dry mix. A typical doughnut batter mix could contain 100% of the flour, 40% sugar and 13% non-emulsified shortening. Other possible ingredients are defatted soy flour, skim milk solids, potato starch and dried egg yolk. Lecithin is sometimes used to assist in wetting the dry mix while monoglycerides are added to make a softer product. Otherwise, emulsifiers are avoided as they increase fat up take.

The protein ingredients reduce fat absorption while potato starch improves shelf life by retaining moisture.

This type of product can be produced by making a dough, rolling it out, cutting it, shaping it and frying it; however, this sort of process is too labour intensive, except for domestic use or small doughnut shops.

Commercial bakeries and doughnut shops normally use a system that extrudes a batter from a reservoir directly into the deep fat fryer (Figure 5). In such a system the rheology of the batter is vitally important. The batter must flow and spread as needed. The major influence on the batter viscosity is the water content, which is around 70% of the flour or around 40% of the dry mix. Various gums are sometimes added to the dry mix to bind water, reduce fat absorption, and control the viscosity. Examples of these gums are guar gum, locust bean gum and carboxymethyl cellulose.

The leavening system is crucial to the success of this product. Typically, 1.5% of sodium bicarbonate and 2.1% sodium acid pyrophosphate (SAPP), both on a flour basis, are used. The SAPP should be a fast acting grade such as SAPP-37 or SAPP-43 so that it acts as fast as possible. The doughnut will drop into the hot fat at 190°C. When leavening takes place the now reduced density doughnut will rise to the surface and the first side will begin to fry. Then the doughnut is flipped over and the second side is fried. In a plant bakery this could take place on an automated production line. During the initial frying the starch has gelatinised and the proteins have been denatured so the

Figure 5 *Automated doughnut making system*

diameter of the doughnut is fixed. The second stage expansion then tends to be into the hole. If the batter is not right the product will "ball up" and the hole fills up.

The progress of cooking inside a doughnut can be investigated by cutting through the doughnut horizontally. If the heat has not reached the innermost part of the doughnut it will be dense, gummy and unpleasant to eat. If the doughnut is overcooked the outside will be case hardened. It is possible to detect which side was fried first as the first cooked side will be less porous. Clearly, the hole assists the passage of heat into the product.

8.7.2 Éclairs

Éclairs are an example of what is called, in English, choux pastry. The pastry is specially made so that it can be piped into shape. The process gelatinises the starch so that the paste has the necessary flow properties to be piped into shape.

The word chou(x) means cabbage(s) in French and is supposed to refer to the shaped produced by baking a lump of this pastry. The product is said to have originated in Renaissance Italy.

8.7.2.1 The Process. Two parts water and one part shortening are boiled together, a little salt is added then one part of flour is added with stirring. The flour is likely to be a soft flour. The now gelatinised flour paste is cooled to around 65°C and two parts of whole egg are added gradually. The paste is cooled and, possibly, a solution of ammonium bicarbonate in skimmed milk is added. The product is then piped into shape on a baking sheet covered with baking parchment and baked in a hot oven at around 220°C. The finished product is leavened by steam and the ammonium bicarbonate if any. The finished piece should be crisp throughout and hollow.

The product is then cut open and filled with cream, or creme patissier. These products are often then coated with chocolate or a chocolate flavour coating.

8.7.3 French Crullers

These are doughnuts made from choux pastry. They appear to be an American product. The choux pastry is shaped into a ring on parchment paper. The rings are allowed to set and the parchment is dipped into the fat and the dough is loosened. Alternatively, the dough can be deposited by a specially shaped cutter. The finished product is less dense than other doughnuts and is sometimes glazed.

8.7.4 Soda Bread

Soda bread is a chemically leavened bread made in Ireland. Instead of yeast the bread is raised chemically using sodium bicarbonate. The traditional source of acid is buttermilk, which contains lactic acid. The use of cream of tartar is now more common. This prevents the bread becoming too alkaline.

Soda bread can be made with a weaker flour than yeast raised bread and should be baked at a cooler temperature. The flour should not have too low a Hagberg Falling Number.

REFERENCE

1. Committee on Medical Aspects of Food Policy (COMA) (1984) Diet and Cardiovascular Disease. Report of the panel on Diet in Relation to Cardiovascular Disease. DHSS Report on Health and Social Subjects 28. London HMSO.

CHAPTER 9

Bread-making Experiments

9.1 INTRODUCTION

Bake testing is an important part of any flour testing regime. It remains true that the odd sample of flour that measures well will bake poorly while the converse situation is also true. The reasons for this are usually because there is a problem with the quality of the protein rather than the quantity. It remains much easier to measure the quantity than the quality of proteins present in flour.

Experiments are ultimately a way of asking a question about nature. All science has, in the end, to be based on experiments.

As well as being an introduction to bake testing this chapter is also an opportunity for the student to discover the effect of different ingredients and processes.

If experiments are to provide a valid source of information care must be taken about weighing and measuring. All ingredients must be from the same batch unless the test is to compare different batches of flour. In any tests there must be a control experiment lest a variation in, *e.g.* in the freshness of the yeast, masks a smaller change owing to some other factor. Another important point is that all experiments should be recorded.

These experiments have been devised on the basis that they can be performed in a domestic kitchen and require no specialised equipment except a robust table mixer, *e.g.* a Kenwood Chef. This is not essential as the experiments could be performed with hand mixing. The other piece of specialised equipment is two loaf tins. These again are not essential as the experiments could be carried out using "bottom bread" by placing the bread on a baking tray, but the experiments are easier if two 2 lb (800 g) loaf tins are available. The experiments will produce surplus dough after the two loaf tins have been filled and this excess can be made into a small loaf or some rolls to sustain the experimenters.

The quantities have been worked out so that each batch can be made from one 1.5 kg retail bag of flour. The recipes can obviously be scaled up or down as required.

9.2 HEALTH AND SAFETY

The following health and safety points have to be made: raw flour is not safe to eat; it can contain bacteria that are harmful to health. For this reason it is essential that hands are washed not only before starting work but also afterwards if raw flour has been handled.

As bread is baked at a high temperature, around 230°C, and ovens get hot, not surprisingly anything that has been in an oven and is hot can burn. Appropriate precautions need to be taken to avoid burns and to apply first aid to any burns that do occur. The usual first aid is to place the afflicted part under cold running water for several minutes. This is a useful treatment because skin is a poor conductor of heat so if the heat can be removed rapidly with running water then the damage can be limited. If the burn is severe then medical attention should be sought.

9.3 YIELD

In all experiments the yield of dough needs to be recorded. Bread production is a competitive business and small variations in yield are important.

9.4 LOAF TESTING

Having made the loaves they have to be tested. When the loaves have cooled they can be measured and assessed. As bread is a perishable product it is convenient to photograph or otherwise make a record of the loaves produced to allow comparisons with samples made on a previous occasion. Digital photography is particularly convenient for this application. An alternative way of recording the amount that a loaf has risen is to photocopy a slice of bread with a ruler to allow the dimensions to be compared with future slices.

Arguably, the appearance, particularly the degree of expansion of a loaf, is not especially important as bread is sold by weight. In practice consumers prefer well-risen and well-shaped loaves.

The first step in recording and assessing the loaves is to record their weight; then the loaves can be assessed for appearance. One factor to look for is the presence of some "oven spring" (see Figures 1–3). This is expansion of the loaf in the oven, which can be detected by stretch marks

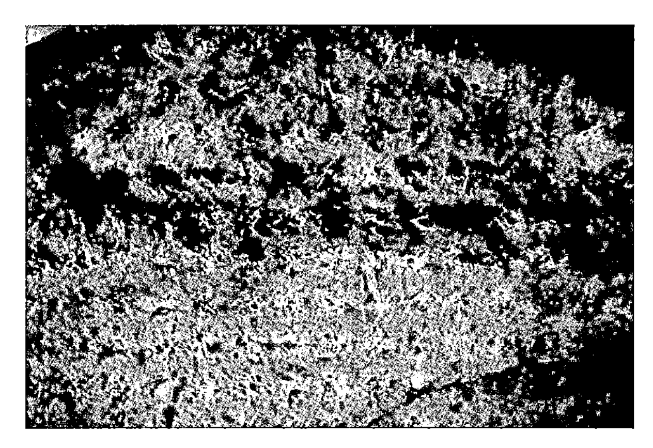

Figure 1 *This loaf has a good lift but the dough has held together*

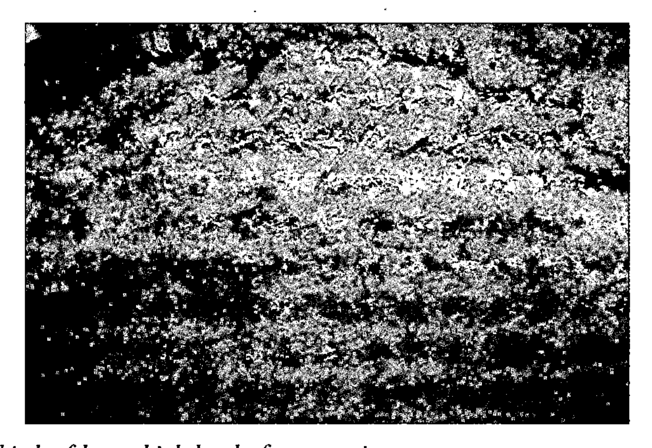

Figure 2 *This loaf has a high level of oven spring*

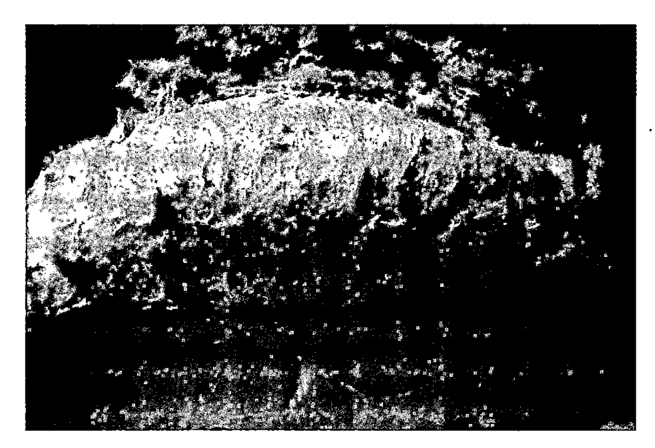

Figure 3 *This loaf was placed in the oven without proving, which has caused very excessive oven spring by its rapid expansion, distorting the loaf*

around the loaf. An excessive amount of oven spring is usually a sign of a lack of proving in the tin.

9.4.1 Tasting

It is impossible to do very much in the food industry without being involved in taste tests. Taste testing is done in several ways, but two common approaches are either to rely on an expert panel or to use a consumer panel and statistical analysis.

Expert panels can give rapid answers. Their disadvantages are that they become so expert that they are sensitive to small differences that the consumer would ignore and that they depend on the palate of the experts. If the experts can not detect a taste, which can happen for genetic reasons, then the taste will not be detected.

One way of conducting a statistical test is the triangular test. In this test, tasters are presented with three samples each. Two of the samples are the same, one is different.

9.5 BREAD MAKING

9.5.1 Recipe

- 1.5 kg flour
- 25 g fat (to be lard or hard vegetable fat)
- 25 g salt
- 15 g sugar
- 25 g fresh yeast
- 800 mL water

9.5.1.1 Notes. This recipe will work with the sort of flour sold at retail in the UK as suitable for bread making. The quantity of water added might need adjusting, particularly for wholemeal flours.

Fresh yeast is definitely best but if it is not available dried yeast will have to be used according to the supplier's instructions.

9.5.2 Straight Method

Blend the dry ingredients thoroughly then add the water at around 40 °C and mix to a dough. The dough then needs to be worked until the gluten develops. Development can be detected because the dough will spring back after being depressed by a thumb.

Mixing can be done either by hand or by machine. If a mixer is used then a robust table mixer like a Kenwood Chef is needed, following the instructions provided with the mixer.

9.5.3 Proving

The dough should either be covered and left in a warm place or put into a purpose built prover until the dough has doubled in size. A purpose built prover would have facilities for applying heat and steam to the product.

9.5.4 Knock Back

The dough is then re-kneaded, which is known as knock back. This redistributes any large bubbles in the dough.

9.5.5 Scaling and Dividing

In this stage the dough is weighed and divided into pieces. First the whole dough is weighed and the yield is recorded. Then two 800 g (or 2 lb) pieces are weighed out. These are rolled into a ball and placed in a tin.

9.5.6 Second Proving

The dough in the tins is then proved for a second time until the dough is above the top of the tins.

9.5.7 Baking

The loaves are then baked at 230°C for 30 min. The loaves can be tested by turning them out of the tins and striking them on the bottom with a knuckle. A cooked loaf will sound hollow when struck.

9.6 SPONGE BATTER

Mix 1 kg of the flour with the salt and fat, then mix the remainder of the flour with the sugar and yeast, and mix in the water to produce a batter. Set the batter on one side until the surface is covered with bubbles then blend in the dry ingredients to produce a dough. The dough must then be kneaded either by machine or by hand until it develops.

9.6.1 Proving

The dough is then covered and proved in a warm place or put in a prover until it has doubled in size.

9.7 VARIATIONS TO THE RECIPE

Whenever a recipe variation is carried out a control sample needs to be made so that the differences if any need to be assessed.

9.7.1 Variation 1: Compare the Effect of Leaving out the Sugar

Using the bulk fermentation method produce a control sample to the standard recipe and a test sample without adding sugar. Compare the two batches to see if adding sugar speeds up the process, checking the effects on yield, loaf shape, loaf volume and product taste.

Repeat the exercise using the sponge batter method. Are the results the same?

On the basis of these tests it should be possible for the experimenter to decide whether it is worthwhile to add sugar or not.

9.7.2 Variation 2: Compare the Effect of Using Vegetable Oil Instead of Hard Fat

Employing any method compare the effect of using the same weight of a vegetable oil such as rape or soy bean oil compared with a control sample made with a hard fat. In this experiment it would be worth comparing the shelf life of the control and the sample.

9.7.3 Variation 3: Compare the Effect of Using No Fat Instead of Hard Fat

In this variation the sample has no fat added while the control is the standard recipe. Either the bulk fermentation or the sponge batter method could be used for this test but the same method should be used as was used in variation 2. Similarly, the shelf life of the bread should be examined.

If sufficient facilities are available then the trials of variations 2 and 3 can be combined as one trial with two samples and one control.

9.7.4 Variation 4: Leave out the Salt

In this variation the sample has no salt added. Any method of bread making can be used as long, obviously, as the sample and the control are made by the same method. Occasionally, this particular error does happen in bakeries. The results of this variation should indicate why salt is added to bread!

9.7.5 Variation 5: Proving in the Sponge Batter Method

This experiment is aimed at determining whether there is a need to prove the dough after the sponge and batter have been mixed together but before the dough is loaded in to the tins. The control batch then is made as per the instructions earlier in this chapter while the sample is made with the dough being weighed into the tins without standing after mixing. In both cases the loaves should be proved before baking.

In this experiment the time to make the standard and the sample should be compared.

9.7.6 Variation 6: Hand Mixing vs. Machine Mixing

To perform a fair test the experimenter will need to practice hand mixing one or more batches before the experiment is carried out. Experienced hand mixers can usually mix a dough as fast as a machine.

To hand mix a dough the hands including under the nails need to be clean and free of cuts and sticking plasters. There are two stages in working on the dough, the first stage is to mix the ingredients together, then the dough must be kneaded until it has developed. Kneading involves stretching and folding the dough, it offers some scope as a stress relieving exercise since the maximum amount of energy needs to be applied to the dough. One of the advantages of hand mixing is that the dough development can be continuously monitored because the feel of the dough changes as it develops.

After some practice the experiment involves timing and comparing a batch produced by hand mixing with a batch produced by machine mixing. Is the quality of the end product indistinguishable?

9.7.7 Variation 7: Comparison of Two Different Flours

An obvious use of bake testing is to compare the performance of two different batches of flour. In this exercise it is suggested that the control batch is a bread flour while the test should be a non-bread-making soft flour such as an English plain flour.

Any method can be used but bulk fermentation is the usual choice. This experiment should give definite differences between the sample and the control.

9.7.8 Variation 8: Testing Different Levels of Water Addition

This variation attempts to address a real problem in determining the optimum level of water addition. Some flour samples need more water added than others.

The experiment uses the bulk fermentation method, making a control sample to the standard recipe and two others, one with 900 mL the other with 1 L of water. On comparing the products it should be apparent which is the best level to use for the flour sample. This variation should be done using a white bread flour.

9.7.9 Variation 9: Wholemeal Flours

Wholemeal bread flour behaves in a slightly different way to white flour so this variation is to work with wholemeal bread flour. In particular, wholemeal flour has a higher water absorption than white flour but no flour treatment is used.

The three levels of water addition used in variation 8, *i.e.* 800 mL, 900 mL and 1 L, should be tried. As a result of this experiment it should be apparent which is the optimum level for this batch of flour. Experimenters are warned that wholemeal flour develops more slowly and will never rise as much as white flour.

9.8 REPORT WRITING

Anyone who has carried out all the experiments above should now have a fair practical knowledge of small-scale simple bread making. The next stage is to produce a report covering the experiments carried out, the results obtained and some conclusions.

CHAPTER 10

The Future

10.1 GENERAL OUTLOOK

What does the future hold for baked products and their manufacture? This is a difficult question to answer and many attempts to predict the future have failed miserably.

The surest way to make a false prediction is to assume that current trends will continue unabated. Mathematically this is known as extrapolation. It is said that the corporation of the City of London examined the trends in traffic in the city in the year 1900. The growth in road traffic had been rapid. At the time of course almost all road traffic was horse drawn, which led to problems with the disposal of manure. If the trend continued it appeared that the manure problem would grow at such a rate that by the year 1960 the city would be 6 feet under manure. Although traffic continued to grow, road transport switched to the internal combustion engine, which ended the manure problem.

Similarly, in the 1960s it was said that one of the problems of the future would be the vast amount of leisure that the population would enjoy. As of the time of writing this leisure problem has failed to appear!

Applying the extrapolation method to the baking industry some years ago the prediction would have been the disappearance of the small baker's shop with a higher proportion of bread being made in large plant bakeries. While small bakeries have continued to close, the total triumph of the plant bakery has not yet happened. Several changes have altered the trend. New technology such as ADD methods have made life easier for the small bakery. Some small bakers have given up the unequal struggle to compete on price with supermarkets and have charged higher prices for a higher quality handmade product. The demand for filled rolls has also produced a useful source of profit for small bakeries. Another competitor for the plant bakeries is the supermarket in-store bakery. These tend to use spiral mixers and suitable improvers to offer the busy supermarket shopper fresh bread. Some consumers have made a move to homemade bread. A factor in some cases is the appearance of automatic home bread-making machines. These machines, which only require measured quantities of ingredients to be loaded and the machine started, are taking a certain amount of trade from bakeries. Thus technology has brought the wheel full circle to a point where many more households are baking their own bread.

In view of the above it is very difficult to predict the future; however, here is an attempt. It is reasonable to suppose that the development of biotechnology will lead to still more enzymes being used in baked products. This could lead to a rapid biochemical system that would replace the Chorleywood bread process with a rapid enzyme-based dough development system.

Another trend that is likely to affect the bakery business is the move to healthier food. This could lead to an increase in products based on wholemeal flour as well as products based on sugar replacers rather than sugar. Special sugar-free bakery products are currently made particularly for the benefit of diabetics.

Another special dietary area is the production of gluten-free bread for those who suffer from coeliac disease. As this is a problem of old age and the population is ageing, demand for this sort of product is increasing.

Yet another trend that is likely to increase is the greater range of products available. Increased travel has produced a demand for more exotic products.

An unknown area is the possibility of new domestic appliances equivalent to the breadmaker. Perhaps an appliance manufacturer is considering a home puff pastry maker.

10.2 DIETARY TRENDS

Dietary trend can be divided into two different classes: trends in what the general population consumes and the trend in reducing diets. In terms of food consumption, the consumption of bread has fallen over the years. Possibly this could reflect the shift of the population from physical to sedentary work.

Reducing diets seem to come and go. At the time of writing the popularity of the Atkins diet seems to be decreasing. The Atkins diet is a low carbohydrate diet that is not conducive to the sales of baked products; however, the Atkins diet seems to be being superseded by diets based on the principle of a low glycemic index. This is unlikely to increase the sale of baked goods but is less antipathetic to them. A new diet based on bread has appeared – and so the wheel may go full circle.

Glossary

ADD: Activated dough development was a no time dough process that relied on L-cysteine and potassium bromate. Not used since potassium bromate has been struck off.
Bulk Fermentation Process: Old-fashioned way of making bread where all the ingredients are fermented together.
Chorleywood Bread Process: Modern method of baking bread using a very high powered mixer and ascorbic acid to make a no time dough.
Creaming: Beating two ingredients together.
Divider: Machine that separates dough into pieces of equal size.
Dockering: Making holes in biscuit dough to assist drying.
Emulsifier: Substance that assists the dispersion of two immiscible substances.
High Ratio Flour: Applied to cake flour that can absorb more than its own weight of sugar and water.
Improver: Substance added to dough to improve its performance.
Knocking Back: Mixing an expanded dough to redistribute the gas bubbles.
Ostwald Ripening: The tendency of crystals to change with time. In response to small temperature variations the smallest crystals redissolve and the largest crystals grow.
Oven Spring: Expansion of product in the oven.
Peel: A wooden blade on a long handle used for removing bread from an oven.
Panning: Putting bread in tins.
Prover: Cabinet where dough can be kept under controlled temperature and humidity to ferment.
Proving or Fermenting: Allowing the yeast to act on the dough.
Retarding: Holding dough at 2–4°C and 80% relative humidity, where it can be held for up to 24 hours before use.
Scaling: Weighing out dough.
Straight Dough: Any method of making bread where all the ingredients are mixed together.

Bibliography

Proteins in Food Processing, ed. R. Y. Yada, Woodhead Publishing, Cambridge, 2004.

Technology of Cereals, Woodhead Publishing, Cambridge, 2001.

D. Manley, *Technology of Biscuits, Crackers and Cookies*, (3rd Edition) Woodhead Publishing, Cambridge, 2000.

Encyclopedia of Food Science and Technology, ed. F. J. Francis, volumes 1–4, John Wiley & Son, 1999.

Cereal Processing Technology, ed. G. Owens, Woodhead Publishing, Cambridge, 2001.

S. Cauvain and L. Young, *Baking Problems Solved*, Woodhead Publishing, Cambridge, 2001.

R. S. Igoe and Y. H. Hui, *Dictionary of Food Ingredients* (4th Edition), Springer-Verlag.

Biscuit Cookie and Cracker Manufacturing Manual 1 Ingredients, Woodhead Publishing, Cambridge, 1998.

Biscuit Cookie and Cracker Manufacturing Manual 2 Biscuit Doughs, Woodhead Publishing, Cambridge, 1998.

Biscuit Cookie and Cracker Manufacturing Manual 3 Biscuit Dough Piece Forming, Woodhead Publishing, Cambridge, 1998.

Biscuit Cookie and Cracker Manufacturing Manual 4 Baking and Cooling of Biscuits Woodhead Publishing, Cambridge, 1998.

P. C. Morris and J. H. Bryce, *Cereal Biotechnology*, Woodhead Publishing, Cambridge, 2002.

R. Guy, *Extrusion Cooking – Technologies and Applications*, Woodhead Publishing, Cambridge, 2001.

Subject Index